Ecological Studies, Vol. 170

Analysis and Synthesis

Edited by

I.T. Baldwin, Jena, Germany
M.M. Caldwell, Logan, USA
G. Heldmaier, Marburg, Germany
R.B. Jackson, Durham, USA
O.L. Lange, Würzburg, Germany
H.A. Mooney, Stanford, USA
E.-D. Schulze, Jena, Germany
U. Sommer, Kiel, Germany

Ecological Studies

Volumes published since 1998 are listed at the end of this book.

Springer

Berlin
Heidelberg
New York
Hong Kong
London
Milan
Paris
Tokyo

H. Sandermann (Ed.)

Molecular Ecotoxicology of Plants

With 42 Figures, 8 in Color, and 9 Tables

 Springer

Prof. Dr. Heinrich Sandermann
GSF-Forschungszentrum für Umwelt und Gesundheit GmbH
Institut für Biochemische Pflanzenpathologie
Ingolstädter Landstraße 1
85764 Neuherberg
Germany

ISSN 0070-8356
ISBN 3-540-00952-3 Springer-Verlag Berlin Heidelberg New York

Library of Congress Cataloging-in-Publication Data

Molecular ecotoxicology of plants / H. Sandermann (ed.)
    p. cm. – (Ecological studies, ISSN 0070-8356 ; v. 170)
    Includes bibliographical references and index.
    ISBN 3-540-00952-3 (hardcover : alk. paper)
    1. Plant molecular biology. 2. Plants, Effect of stress on. 3. Plant ecophysiology. I.
Sandermann, Heinrich. II. Series.

QK728.M6 2003
572.8'--dc21   2003054788

Springer-Verlag Berlin Heidelberg New York
a company of BertelsmannSpringer Science+Business Media GmbH

http://www.springer.de

© Springer-Verlag Berlin Heidelberg 2004

Printed in Germany

Production: Friedmut Kröner, 69115 Heidelberg, Germany
Cover design: design & production GmbH, Heidelberg
Typesetting: Kröner, 69115 Heidelberg, Germany

31/3150 YK – 5 4 3 2 1 0 – Printed on acid free paper

Preface

The idea for this book publication came from a previous publication in Ecological Studies (Sandermann et al. 1997). In that book it was stated that the best way to study forest decline, or forest ecosystems generally, is to work along an experimental hierarchy where the plant is treated as an individual, as part of its microenvironment and finally as part of an ecosystem. One method of ecological research therefore is to scale up from the individual plant to the ecosystem by studying organismic ecophysiology and ecosystem flows. This approach is taken in several volumes of Ecological Studies. A second approach (the one taken here) is to concentrate on early plant responses and stress-induced transcripts, while making use of the present plant genome projects. This latter approach has to include *Arabidopsis thaliana*, whose genome project is by far the most advanced.

Molecular plant stress responses, about which numerous book publications exist, are not studied as such. The aim is to relate molecular stress responses and signalling pathways in the plant to environmental interactions. On one hand, the environment delivers input parameters such as chemical or physical stressors, or pathogen elicitors. This is followed by signalling and stress response events. The plant is thereby biochemically changed so that different output parameters are delivered to subsequent environmental interactions. This concerns in particular a changed disposition for pathogen attack, senescence or other stress episodes.

The fitness of the individual plant and the dynamics of the plant population will be changed. This basic idea can be summarized as follows:

Input from the environment	$\rightarrow$	Signalling and stress responses in the plant	$\rightarrow$	Output into the environment

At present, it is often not clear whether molecular plant stress responses observed in the laboratory actually occur in field situations, where multiple stress exposure is typical. The initial discussion of this point and the inclusion of some of the relatively few available field data set the present book apart from the many book publications on mechanisms of plant signalling and stress responses. Of course, the latter are needed, and also included here, in order to define the basic rules. The present book is selective in that only a limited set of case studies is presented. It should be noted that much more ambitious projects such as clarification of the environmental significance of each of the about 25,000 genes of *Arabidopsis thaliana* are under way and have a good chance of success because of powerful high-throughput techniques (Jackson et al. 2002).

The core of the present group of authors has contributed to a plant stress symposium at the 16th International Botanical Congress, St. Louis, August 1999 (Sandermann and Pell 1999). Special thanks are due to the co-organizer of that symposium, Professor Eva J. Pell, Pennsylvania State University, State College, USA. The Senior Editor of Ecological Studies, Dr. Ian T. Baldwin, Max Planck Institute for Chemical Ecology, Jena, made many helpful suggestions for improvement. The technical support by Drs. D. Czeschlik and A. Schlitzberger of the Springer-Verlag Heidelberg and the cooperation of all contributors are also gratefully acknowledged. It is hoped that this book can help to introduce molecular techniques into ecological research, so that a new level of understanding of plant and ecosystem health can be achieved.

Neuherberg, May 2003   *Heinrich Sandermann*

References

Sandermann H, Wellburn AR, Heath RL (1997) Forest decline and ozone. Ecological studies 127. Springer, Berlin Heidelberg New York

Sandermann H, Pell E (1999) Abstract volume, 16th International Botanical Congress, Symposium 13.3, Plant stress response: plasticity for natural and anthropogenic stressors, St Louis, MO, 168 pp

Jackson RB, Linder CR, Lynch M, Purugsanan M, Somerville S, Thayer SS (2002) Linking molecular insight and ecological research. Trends Ecol Evol 17:409–414

Contents

4   **Ethylene and Jasmonate as Regulators
     of Cell Death in Disease Resistance**   75
     C. Langebartels and J. Kangasjärvi

Contributors

AARTS M.

Laboratory of Genetics, Department of Plant Sciences, Wageningen University, Postbus 16, 6700 AA Wageningen, The Netherlands, e-mail: mark.aarts@wur.nl

DIXON D.

Crop Protection Group, University of Durham, Durham, DH1 3LE, UK, e-mail: d.p.dixon@durham.ac.uk

DURNER J.

GSF–Forschungszentrum für Umwelt und Gesundheit GmbH, Institut für Biochemische Pflanzenpathologie, Ingolstädter Landstraße 1, D-85764 Neuherberg, Germany, e-mail: durner@gsf.de

EDWARDS R.

Crop Protection Group, University of Durham, Durham, DH1 3LE, UK, e-mail: robert.edwards@durham.ac.uk

ERNST D.

GSF–Forschungszentrum für Umwelt und Gesundheit GmbH, Institut für Biochemische Pflanzenpathologie, Ingolstädter Landstraße 1, D-85764 Neuherberg, Germany, e-mail: ernst@gsf.de

GLICK B. R.

Department of Biology, University of Waterloo, Waterloo, Ontario, Canada N2L 3G1, e-mail: glick@sciborg.uwaterloo.ca

KANGASJÄRVI J.

Department of Biosciences, University of Helsinki, 100014 Helsinki, Finland, e-mail: Jaakko.Kangasjarvi@helsinki.fi

LANGEBARTELS C.

GSF–Forschungszentrum für Umwelt und Gesundheit GmbH,
Institut für Biochemische Pflanzenpathologie, Ingolstädter Landstraße 1,
D-85764 Neuherberg, Germany, e-mail: langebartels@gsf.de

MATYSSEK R.

Technische Universität München, Lehrstuhl für Forstbotanik,
Hohenbachernstraße 22, 85354 Freising-Weihenstephan, Germany,
e-mail: matyssek@bot.forst.uni-muenchen.de

MÉTRAUX J.-P.

Département de Biologie, Université de Fribourg, Rue A.-Gockel 3,
1700 Fribourg, Switzerland, e-mail: jean-pierre.metraux@unifr.ch

MITTLER R.

Department of Biochemistry, University of Nevada, Fleischmann Agriculture
307/Mailstop 200, Reno, Nevada 89557, USA, e-mail: rmittler@unr.edu

SANDERMANN H.

GSF–Forschungszentrum für Umwelt und Gesundheit GmbH,
Institut für Biochemische Pflanzenpathologie, Ingolstädter Landstraße 1,
D-85764 Neuherberg, Germany, e-mail: sandermann@gsf.de

VAN LOON L. C.

Faculty of Biology, Section Phytopathology, Utrecht University,
Post Office Box 800.84, 3508 TB Utrecht, The Netherlands,
e-mail: l.c.vanloon@bio.uu.nl

ZILINSKAS B. A.

Department of Plant Biology and Pathology, Rutgers University,
59 Dudley Road, New Brunswick, New Jersey 08901, USA,
e-mail: zilinskas@aesop.rutgers.edu

1 Molecular Ecotoxicology: From Man-Made Pollutants to Multiple Environmental Stresses

H. Sandermann

1.1 Overview

All life forms depend on plants as primary producers of food and feed and of substrates for decomposition. In the mid-1960s, it was realized that the productivity and biodiversity of whole ecosystems, including their vegetation, were threatened by man-made pollutants. Later on, other natural and man-made influences, such as climatic extremes, pathogens, herbivores or air pollution, were also studied as environmental stressors. The existence of highly effective plant acclimation and stress defense mechanisms is demonstrated by the occurrence of plants on extreme environmental sites and by the resistance of plants to most pathogens. Plants can also detoxify man-made chemicals, as shown by the selective action of many herbicides that kill weeds without inhibiting the growth of crop plants.

The defense mechanisms of plants are generally of a multipurpose nature. For example, the enzyme families detoxifying man-made chemicals can also protect against toxic metabolic by-products and natural toxins. Plant substances protecting against UV-B can also be antioxidants and antibiotics. Plant genes involved in pathogen defense can also be important for defense against air pollutants, and genes of normal protein assembly may be involved in protection against elevated temperature. Such cross-protective mechanisms are in many cases still tentative, but the recently completed plant genome projects and current array methods, as well as genetic overexpression or knock-out, are likely to identify the signal chains and molecular response programs involved. The purpose of this book is to provide an introduction into the newly developing research field of molecular ecotoxicology and to present a number of case studies.

Ecological Studies, Vol. 170
H. Sandermann (Ed.)
Molecular Ecotoxicology of Plants
© Springer-Verlag Berlin Heidelberg 2004

1.2 Definitions of Ecotoxicology

Ecotoxicology is defined here as the science of the occurrence, transformation, bioavailability or exposure and the mode of action of physical, chemical and biological environmental stress factors. Ecotoxic effects need to be determined for individual plant species, for populations and biocoenoses and, finally, for whole ecosystems.

In the early 1960s, scientists first became aware of the detrimental effects of man-made pollutants (Farb 1965). The famous book, *Silent Spring*, by Rachel Carson (1962) had made a vast public aware of the possible toxic side effects that chlorinated insecticides and herbicides exerted on non-target organisms in the environment, the emphasis being on bird populations. The widespread use of insecticides had led to the emergence of resistant insect populations (Carson 1962; Farb 1965). Later on, pathogens resistant to fungicides appeared in agricultural ecosystems, and the spontaneous herbicide resistance of weeds also became a widespread problem (LeBaron and Gressel 1982). In the case of atrazine, it could be demonstrated that a point mutation in the chloroplast psb A gene led to an amino-acid change that made the chloroplast target protein insensitive to atrazine (Hirschberg and McIntosh 1983). By this mechanism, the empirical observation of herbicide resistance was explained at the molecular level. Later research clarified more mechanisms of herbicide resistance and its spread in weed populations (Powles and Holtum 1994). These various results document an early thorough case study of molecular ecotoxicology of plants, molecular mechanisms being successfully scaled up to the ecosystem level. Fungal subspecies with spontaneous fungicide resistance and insect subspecies with spontaneous insecticide resistance have also increased dramatically. These unintended developments can now be useful for the ecological assessment of transgenic plants with pesticidal traits (see Chapter 8).

Carson described many examples of undesirable ecotoxic side effects of pesticides, but the term "ecotoxicology" was used only subsequently when large-scale scientific research and legislative efforts were initiated. For example, the USA established the Toxic Substances Control Act (1976), and Germany banned DDT in 1972 and established a Chemicals Law in 1980. Many test procedures were developed to determine the ecotoxic character of anthropogenic pollutants. Bioaccumulation and biodegradation processes became understood, and today predicted environmental concentrations and predicted no-effect levels of chemicals are known at least for short-term tests and defined test species (including algae and higher plants).

Concerning plants, the modes of action and of selectivity of herbicides have been studied in particular detail. The target proteins and their genes often are unique for plants, such as in photosynthesis and branched-chain or aromatic amino acid biosynthesis. The absence of these targets in humans

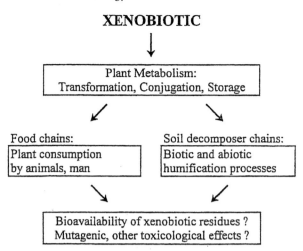

XENOBIOTIC

↓

| Plant Metabolism: Transformation, Conjugation, Storage |

Food chains: | Plant consumption by animals, man |

Soil decomposer chains: | Biotic and abiotic humification processes |

| Bioavailability of xenobiotic residues ? Mutagenic, other toxicological effects ? |

Fig. 1.1. Schematic summary of the toxicological potential of xenobiotic plant metabolites. (Sandermann 1994)

and animals could secure a low level of toxic side effects. Plants have been found to resemble animals in having similar metabolic enzymes of transformation, conjugation and compartmentalization. This similarity has led to the "green liver" concept and to the summary of ecotoxicological significance of plant pesticidal residues that is shown in Fig. 1.1. As in animals, plant metabolism can result in detoxification or bioactivation. Remarkably, similar enzyme classes were found to metabolize xenobiotics and natural substrates (Table 1.1; Sandermann 1992). These early observations have been confirmed and much extended by the detection of large homologous gene families in *Arabidopsis thaliana* (The *Arabidopsis* Genome Initiative 2000). This example already illustrates the enormous gain in information by present genome projects. Enzyme families, e.g., of cytochromes P450 and glucosyltransferases, are responsible for xenobiotic metabolism, but also for regulation of concentrations of free signalling molecules such as jasmonic and salicylic acids, auxins and cytokinins. Mutagenic effects of environmental stressors have recently been visualized using the recombination of a split and inactive β-glucuronidase marker gene to yield active enzyme producing a blue color. In this way, toxic effects of realistic UV-B (Ries et al. 2000) and heavy metal ions (Kovalchuk et al. 2001) were demonstrated in a dose-dependent manner.

Ecotoxicology was initially developed as a scientific discipline limited to studying the biological consequences of anthropogenic chemicals in the environment (e.g., Korte 1980). A broader definition has included any physical (heat, radiation), chemical or biological factor that creates a potential source of pollution (Ramade 1979). This extended definition still limited the scope of ecotoxicology to man-made pollutants, whereas the present definition also includes natural substances as well as physical parameters. Some of these stressors are listed in Table 1.2. In the typical situation of low-level exposure to man-made pollutants, their ecotoxic effects may be masked by those of nat-

Table 1.1. Plant enzyme classes for either xenobiotic or natural substrates. (Sandermann 1992)

Enzyme class	Xenobiotic substrates	Natural substrates
Cytochromes P450	4-Chloro-*N*-methyl-aniline	Cinnamic acid, pterocarpans
Glutathione *S*-transferases	Fluorodifen, alachlor, atrazine	Cinnamic acid
Carboxylesterases	Diethylhexylphthalate	Lipids, acetylcholine
O-Glucosyltransferases	Chlorinated phenols	Flavonoids, coniferyl alcohol
O-Malonyltransferases	β-D-Glucosides of penta chloro-phenol and of 4-hydroxy-2,5-di-chloro-phenoxyacetic acid	β-D-Glucosides of flavonoids and isoflavonoids
N-Glucosyltransferases	Chlorinated anilines, metribuzin	Nicotinic acid
N-Malonyltransferases	Chlorinated anilines	1-Aminocyclopropyl carbo-xylic acid, D-amino acids, anthranilic acid

Table 1.2. Selected environmental stress factors

Abiotic	Biotic
Air pollutants	Allelopathy
Allelochemicals	Herbivores
Fire	Microbial pathogens
Heavy metals	Viral pathogens
Salinity	Nematodes
Mechanical stress	Plant/plant competition
Nutrient deficiencies	Pollinators
Pesticides	Symbiont availability
Radiation	
Shading	
Temperature extremes	
Toxins	
Water deficit or excess	

ural stressors. In any case, plants are seen as targets as well as producers of environmental stress, both aspects being involved in determining plant fitness.

1.3 Environmental Stress

It has long been known that many stress and selection factors exist in the environment. Ernst Haeckel, in 1866, first used the term "ecology" in the sense of the "economy of nature" and the "science of all interactions between organisms and their environment." The ability of organisms to adapt to their environmental conditions seemed important to withstand parasitism and to fight for survival (Haeckel 1911). Exposure to the natural environment thus was seen already to constitute stress. It is still a difficult problem to judge the significance of superimposed anthropogenic stress in ecosystems. The book, *Silent Spring*, (Carson 1962) presented the following example of natural ecotoxic stress. Roses in a park in a city in Holland suffered from heavy infestations by tiny nematode worms. Instead of chemical treatments, marigolds were planted among the roses. The marigolds released an excretion from their roots that killed the soil nematodes and allowed the roses to flourish. This story is based on the allelopathic principle first described by H. Molisch in 1937. Allelopathy has been defined as chemical competition between plants (Harborne 1993). Another natural ecotoxic principle, that of phytoalexins, was first described by Müller and Börger in 1941. Phytoalexins are plant compounds that are toxic for fungi. These initial and later discoveries were well summarized in Harborne's *Introduction to Ecological Chemistry* (1993).

Plants employ the two mentioned and many other chemical mechanisms to defend themselves against a broad range of biotic and abiotic stressors. The active principles of plants can be seen as "natural pesticides" that can be constitutive (Wittstock and Gershenzon 2002) or induced (Ebel 1986; Harborne 1993). As with man-made pesticides, there are natural herbicides (allelochemicals), fungicides (phytoalexins), antibiotics, insecticides, acaricides, etc., all of which can also exert non-target effects on plants as well as animals and humans. Natural pesticides can be acutely toxic (e.g., alkaloids), estrogenic (e.g., soybean isoflavonoids) or neurotoxic (e.g., hallucinogens). Natural plant compounds thus open a broad field of toxicology and ecotoxicology with basically the same questions as in the case of man-made chemicals. A greatly simplified sequence to analyze the stress exposure of plants is outlined in Fig. 1.2.

Many physical and environmental stressors act on plants as mimics of biotic elicitors, which were first characterized in the case of cell wall fractions of the plant-pathogenic oomycete *Phytophthora megasperma*. This elicitor was defined by its ability to induce phytoalexins in soybean plants and cell cultures. Subsequently, the elicitor was recognized also to induce many other

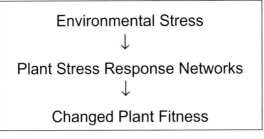

Fig. 1.2. Logical sequence to study the stress exposure of plants

defense responses (Ebel 1986). Today, *Phythophthora megasperma* is known to produce specific oligosaccharide and peptide elicitors that are recognized with high affinity by receptor proteins. Signal chains and the expression of numerous defense genes are activated (Ebel and Mithöfer 1998). The elicitor concept was initially developed as part of plant/pathogen interactions. However, it was realized at an early stage that plants are also responsive to abiotic environmental elicitors, such as heavy metal ions (Moesta and Grisebach 1980). The latter are restricted to special growth locations of plants, but more recently ozone, a ubiquitous air pollutant that regularly exceeds limit values, has also been recognized to act as a powerful abiotic elicitor (Sandermann 1996; Pell et al. 1997; Sandermann et al. 1998; Rao and Davis 2001). As summarized in Fig. 1.3, ozone can induce five major known defense responses of plants against pathogens. Ozone therefore acts as an inducer that is able to activate plant genetic and biochemical programs originally developed for another environmental stress (in this case fungal attack).

The signalling chains as well as the responses at the transcript, protein and metabolite levels partially overlap and are thus mutually connected, as depicted in the Venn diagram of Fig. 1.4. Part of the signalling and response networks are typically shared with normal development, whereas other parts have apparent stress specificity. Figure 1.4 shows two-dimensional overlaps, and the conceptual extension to multidimensional overlap is also sketched. Plants experience many developmental phases and are typically exposed to multiple environmental stress factors. Much of the current plant stress research can be interpreted by such overlap diagrams, for example, ozone actions overlapping those of pathogen attack (Sandermann 1996) or herbivory responses overlapping with those of wounding and pathogen attack (Kessler and Baldwin 2002). This type of diagram is useful to analyze overlapping complex gene induction patterns, as illustrated by the array transcript analysis of the effects of jasmonic acid, salicylic acid and ethylene on *Arabidopsis*. The overlaps contained between 1 and 55 transcripts for induction and between 0 and 28 transcripts for repression (Schenk et al. 2000). Recent analyses of transcripts induced in *Arabidopsis* and tobacco by oxidative stress used a non-interacting tart diagram and cluster diagrams, respectively (Desikan et al. 2001; Vranova et al. 2002). While the Venn diagram

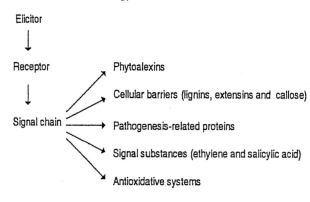

Fig. 1.3. General scheme for the induction of plant defense systems by fungal elicitor and by ozone. (Taken from Sandermann et al. 1998)

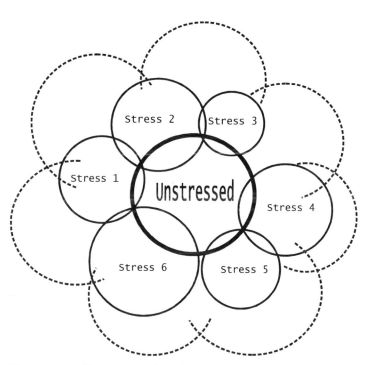

Fig. 1.4. Venn diagram to analyze the overlap of plant stress responses with processes of normal plant development (shown as *unstressed plant*). There may be overlaps between unstressed and stressed plants and also between the individual stressors. There may even be triple overlap areas. An excellent Venn diagram for transcript analysis can be found in Schenk et al. (2000). However, Venn diagrams are also useful to analyze protein and metabolite patterns, genotypes within a plant population, etc. Plants are typically exposed to multiple stresses, so that the two-dimensional diagram needs to be made multidimensional. This is sketched here, although this extension can really be done only with suitable computer software

appears useful in characterizing functional overlap, functional response networks and dynamic interactions have to be analyzed by more rigorous mathematical approaches. On the other hand, the network character of hydrogen peroxide signalling (Neill et al. 2002) and of pathogen defense (Kunkel and Brooks 2002) have recently been well described without the use of formal diagrams.

Plant stress response programs can have a hierarchical order. For example, heat shock is able to silence the phytoalexin as well as UV-B responses of parsley (Walter 1989) and the defense response of tomato (Kuun et al. 2001). Elicitor treatment silenced the UV-B response of parsley (Logemann and Hahlbrock 2002). Symbiotic rhizobacteria can silence the defense responses induced by elicitors in soybean (Mithöfer 2002). A fungal saponin-detoxifying enzyme could also mediate suppression of plant defenses (Bouarab et al. 2002). On the other hand, ozone could simultaneously induce the UV-B and fungal defense responses in parsley plants. This process of "cross-induction" is depicted in Fig. 1.5 (Eckey-Kaltenbach et al. 1994). Ozone could also induce a heat shock gene (Eckey-Kaltenbach et al. 1996). It is now known that such cross-induction phenomena are widespread in plants, as summarized in Table 1.3. The inducers generally act via signalling chains that involve second messengers such as activated oxygen species, salicylic acid, oxylipins such as jasmonic acid, the classical phytohormones, Ca^{2+} ions and protein kinase cascades. Oxylipins and insect-derived elicitors such as volicitin (Kessler and Baldwin 2002) have been shown to induce emission of volatile secondary plant compounds. Ozone has recently been found to act similarly, even inducing large amounts of the second messenger molecules methylsalicylate (Heiden et al. 1999) and ethylene (Sandermann 1996). This illustrates again the elicitor-like nature of ozone and the mutual mimicry of abiotic and biotic

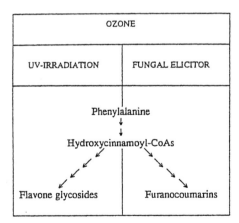

Fig. 1.5. Model for the effects of ozone, UV irradiation and fungal elicitor on the defensive pathways in parsley. (Taken from Eckey-Kaltenbach et al. 1994)

Table 1.3. Cross-induction of plant defense programs

Stress response	Primary inducer	Additional inducer (lead references)
Heat-shock proteins	Elevated temperature	Arsenite, heavy metal ions, development (Nover et al. 1989)
Phytoalexin biosynthesis	Fungal elicitor	Heavy metal ions, UV-C, detergents, ozone, development (Moesta and Grisebach 1980; Ebel 1986; Sandermann 1996)
Pathogenesis-related proteins	Microbial or viral infection	Aspirin, ozone, heavy metal ions, Basta® herbicide, UV-B, development (Bol et al. 1990)
Protease inhibitor	Herbivores	Wounding, development (Koiwa et al. 1997)
Senescence genes and proteins	Development	Ozone, drought, ethylene, pathogens (Quirino et al. 2000)

stress. The rules for cross-induction or for silencing are not well understood at present, and such basic mechanisms of plant molecular ecotoxicology need to be systematically elucidated.

1.4 The Unexpected Side of Ecotoxicology: Induced Resistance

Environmental stressors are not only detrimental to plants fitness, but can also help plants to become more resistant to a second stress episode. For example, a heat shock episode protects against a second heat treatment or against viral infection (Nover et al. 1989). After inoculation with viral pathogens, plants became resistant to a second viral inoculation. Prior treatment with salicylic or 2,6-dichloro-isonicotinic acid also protected plants against subsequent viral infection, as first observed for tobacco mosaic virus (Agrios 1997). The air pollutant ozone was able to induce the five major known defense systems of plants (see Fig. 1.3) and is thus predicted to strengthen plant defense against pathogens in addition to its well-known detrimental effects (Sandermann 1996). Indeed, it has been documented repeatedly that ozone treatment can protect plants against pathogen attack (Manning and von Tiedemann 1995; Langebartels et al. 2001), but in many other cases pathogen attack was enhanced (Manning and von Tiedemann 1995). The various observations of induced resistance lead to a generalized view of environmental stressors that is summarized in Fig. 1.6. Instead of focusing exclusively on acute and chronic toxic effects, the research task also includes induced local and systemic resistance phenomena and their ecological consequences. Changes in metabolic status and disposition of stressed plants and enhanced susceptibility or induced resistance to a second stress factor need to be studied.

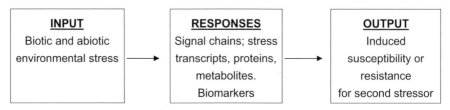

Fig. 1.6. Action of environmental stressors (*input*) on plants (a functional extension of Fig. 1.3)

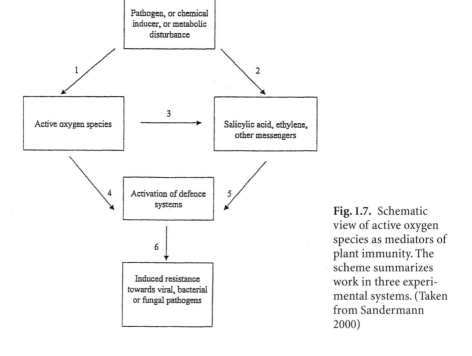

Fig. 1.7. Schematic view of active oxygen species as mediators of plant immunity. The scheme summarizes work in three experimental systems. (Taken from Sandermann 2000)

Stress-inducible plant defense systems are increasingly seen to resemble the innate immunity systems of animals and humans (Cao et al. 2001; Janeway and Medzhitov 2002; Gómez-Gómez and Boller 2002), so that further research may uncover old evolutionary principles of adaptation and survival. One principle of induced resistance that is documented by several case studies involves mediation by active oxygen species, in particular hydrogen peroxide (Sandermann 2000; Fig. 1.7). Hydrogen peroxide may generally act as a systemic signal in plants (Neill et al. 2002).

1.5 Novel Techniques

The presence of multiple rather than single biotic and abiotic stressors is typical for most ecosystems, and powerful methods of molecular analysis have recently been developed to elucidate the complex stress response networks. DNA sequencing and bioinformatics have led to the first completed plant genome project in the case of *Arabidopsis thaliana* and in an assignment of some 10 % of the genome to plant defense and cell death (The *Arabidopsis* Genome Initiative 2000). The rice genome project has led to a somewhat higher percentage (Goff et al. 2002). Novel techniques making use of genomic information have been well summarized (Gibson 2002; Jackson et al. 2002). The transcript patterns of stressed and non-stressed plants are compared by array techniques, an example being shown in Fig. 1.8. Transcript analysis is followed by analysis of the encoded proteins (proteomics and biochemical

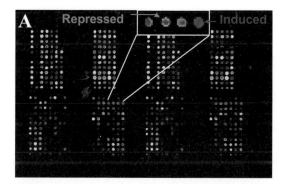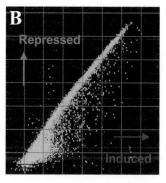

Fig. 1.8A, B. (in color at end of book) DNA microarray analysis of nitric oxide-induced gene expression in *Arabidopsis thaliana*. A Original scan of a DNA microarray hybridized with fluorescent-labeled samples (*green*, untreated control; *red*, sample from plants treated with nitric oxide, *NO*). *Green spots* indicate suppression, and *red spots* induction of the corresponding gene. B Global survey of transcriptional activity determined by a DNA array. The *scatter plot* shows significant gene activation after NO treatment (shift to the *right*). (Data from Jörg Durner, GSF-Forschungszentrum)

functional analysis) as well as by metabolite profiling and functional assignment of the metabolites. The general research task is to correlate the plant genotype with the phenotype and "ecotype." Many plant stress metabolites are volatilized from plants or exuded from roots, so that nearby organisms may also be affected. Excellent case studies have been performed in the case of herbivory, where the following processes were studied (Walling 2000; Kessler and Baldwin 2002): transcript accumulation, emission of volatiles, distinction of effects induced by wounding and by elicitors in insect spit, the role of second messenger molecules and cross-induction of reactions of pathogen defense. One of the many questions raised concerns the ecological significance of volatile plant secondary compounds released in wounding, herbivory or ozone action.

1.6 Scope of the Present Book

The topics in the chapters to follow range from man-made pollutants to multiple environmental stresses. With regard to man-made pollutants, the descriptions of the "green liver" concept (Sandermann 1992, 1996) have been followed by more extensive reviews that have increasingly taken the *Arabidopsis* genome project into account. All major detoxifying enzyme systems have been covered in a book publication (Hatzios 1997), and the biotechnological use of these enzymes and their genes for phytoremediation have been considered (Meagher 2000; Davis et al. 2002; Schäffner et al. 2002). The major metabolic enzyme families for xenobiotics have been reviewed in detail, viz. glycosyltransferases (Vogt and Jones 2000), cytochromes P450 (Werck-Reichhart et al. 2000; Feldmann 2001) and glutathione S-transferases (Edwards et al. 2000). An update of the latter review constitutes the first case study presented in this book (Chap. 2). Subsequent chapters deal with multiple environmental stresses involving activated oxygen species (Chap. 3), second messenger systems (Chaps. 4 and 5), stress-responsive promoters (Chap. 6) and increased fitness by rhizobacteria (Chap. 7). A summary of these chapters is presented in the final Chapter 8 together with some more general topics.

Parts of the subject area of this book have been dealt with in previous book publications. The books by Fritz and Simms (1992), Harborne (1993), Bazzaz (1996), Brunold et al. (1996) and Karban and Baldwin (1997) deal with plant reactions against natural environmental stressors. The book edited by Hock and Elstner (1995) also includes xenobiotics. The quoted books do not make much reference to the molecular level. In contrast, the books edited by Inzé and van Montagu (2002) and Scheel and Wasternack (2002) emphasize molecular aspects of oxidative stress and plant signal transduction, respectively. The present book publication attempts to link molecular and ecological studies from the laboratory to the field levels to the limited extent presently possible. This attempt is in line with a recent review on microarray use in ecology that concludes that molecular methods have not yet arrived in research on ecology and evolution, although their potential is clear (Gibson 2002).

References

Agrios GN (1997) Plant pathology, 4th edn. Academic Press, San Diego

Barker JR, Tingey DT (eds) (1992) Air pollution effects on biodiversity. Van Nostrand Reinhold, New York

Bazzaz FA (1996) Plants in changing environments. Linking physiological, population, and community ecology, Cambridge University Press, Cambridge

Bol JF, Linthorst HJM, Cornelissen BJC (1990) Plant pathogenesis-related proteins induced by virus infection. Annu Rev Phytopathol 28:113–138

Bouarab K, Melton R, Baulcombe D, Osbourn A (2002) A saponin-detoxifying enzyme mediates suppression of plant defences. Nature 418:889–892

Brunold C, Rüegsegger A, Brändle R (eds) (1996) Stress bei Pflanzen. Verlag Paul Haupt, Bern

Cao H, Baldini RL, Rahme LG (2001) Common mechanisms for pathogens of plants and animals. Annu Rev Phytopathol 39:259–284

Carson R (1962) Silent spring. Houghton Mifflin Co, Boston, USA

Davis LC, Castro-Diaz S, Zhang Q, Erickson LE (2002) Benefits of vegetation for soils with organic contaminants. Crit Rev Plant Sci 21:457–491

Desikan R, A-H-Mackerness S, Hancock JT, Neill SJ (2001) Regulation of the *Arabidopsis* transcriptome by oxidative stress. Plant Physiol 127:159–172

Ebel J (1986) Phytoalexin synthesis: the biochemical analysis of the induction process. Annu Rev Phytopathol 24:235–264

Ebel J, Mithöfer A (1998) Early events in the elicitation of plant defence. Planta 206:335–348

Eckey-Kaltenbach H, Ernst D, Heller W, Sandermann H Jr (1994) Biochemical plant responses to ozone. IV. Cross-induction of defensive pathways in parsley (*Petroselinum crispum* L.) plants. Plant Physiol 104:67–74

Eckey-Kaltenbach H, Kiefer E, Grosskopf E, Ernst D, Sandermann H Jr (1996) Differential transcript induction of parsley pathogenesis-related proteins and of a small heat shock protein by ozone and heat shock. Plant Mol Biol 33:343–350

Edwards R, Dixon DP, Walbot V (2000) Plant glutathione S-transferases: enzymes with multiple functions in sickness and in health. Trends Plant Sci 5:193–198

Farb P (1965) Ecology. Life Nature Library, Time-Life International (Nederland) NV, Amsterdam

Feldmann KA (2001) Cytochrome P450s as genes for crop improvement. Curr Opin Plant Biol 4:162–167

Feys BJ, Parker JE (2000) Interplay of signaling pathways in plant disease resistance. TIG 16:449–455

Fritz RS, Simms EL (eds) (1992) Plant resistance to herbivores and pathogens. Ecology, evolution, and genetics. The University of Chicago Press, Chicago

Gibson G (2002) Microarrays in ecology and evolution: a preview. Mol Ecol 11:17–24

Goff SA, Ricke D, Lan TH, Presting G, Wang R et al. (2002) A draft sequence of the rice genome (*Oryza sativa* L. ssp. *japonica*). Science 296:92–100

Gómez-Gómez L, Boller T (2002) Flagellin perception: a paradigm for innate immunity. Trends Plant Sci 7:251–256

Haeckel E (1911) Natürliche Schöpfungsgeschichte. Verlag G Reimer, Berlin pp 793–794

Harborne JB (1993) Introduction to ecological biochemistry, 4th edn. Academic Press, London

Hatzios KK (ed) (1997) Regulation of enzymatic systems detoxifying xenobiotics in plants. Kluwer, Dordrecht

Heiden AC, Hoffmann T, Kahl J, Kley D, Klockow D, Langebartels C, Mehlhorn H, Sandermann H Jr, Schraudner M, Schuh G, Wildt J (1999) Emission of volatile organic compounds from ozone-exposed plants. Ecol Application 9:1160–1167

Hirschberg J, McIntosh I (1983) Molecular basis of herbicide resistance in *Amaranthus hybridus*. Science 222:1346–1349

Hock B, Elstner EF (1995) Schadwirkungen auf Pflanzen. 3rd edn. Spektrum Akademischer Verlag, Heidelberg

Inzé D, Van Montagu M (eds) (2002) Oxidative stress in plants. Taylor and Francis, London

Jackson RB, Linder CR, Lynch M, Purugganan M, Somerville S, Thayer SS (2002) Linking molecular insight and ecological research. Trends Ecol Evol 17:409–414

Janeway CA Jr, Medzhitov R (2002) Innate immune recognition. Annu Rev Immunol 20:197–216

Karban R, Baldwin IT (1997) Induced responses to herbivory. The University of Chicago Press, Chicago

Kazuya I, Tena G, Henry Y, Champion A, Kreis M, Hirt H et al. (2002) Mitogen-activated protein kinase cascades in plants: a new nomenclature. Trends Plant Sci 7:301–308

Kessler A, Baldwin IT (2002) Plant responses to insect herbivory: the emerging molecular analysis. Annu Rev Plant Biol 53:299–328

Koiwa H, Bressau RA, Hasegawa PM (1997) Regulation of protease inhibitors and plant defense. Trends Plant Sci 2:379–384

Korte F (ed) (1980) Ökologische Chemie. Grundlagen und Konzepte für die ökologische Beurteilung von Chemikalien. Georg Thieme Verlag, Stuttgart

Kovalchuk O, Titov V, Hohn B, Kovalchuk I (2001) A sensitive transgenic plant system to detect toxic inorganic compounds in the environment. Nat Biotechnol 19:568–572

Kunkel BN, Brooks DM (2002) Cross talk between signaling pathways in pathogen defense. Curr Opin Plant Biol 5:325–331

Kuun KG, Okole B, Bornman L (2001) Protection of phenylpropanoid metabolism by prior heat treatment in *Lycopersicon esculentum* exposed to *Ralstonia solanacearum*. Plant Physiol Biochem 39:871–880

Langebartels C, Heller W, Führer G, Lippert M, Simons S, Sandermann H Jr (1998) Memory effects in the action of ozone on conifers. Ecotox Environ Safety 41:62–72

LeBaron HM, Gressel J (eds) (1982) Herbicide resistance in plants. Wiley, New York

Logemann E, Hahlbrock K (2002) Crosstalk among stress responses in plants: pathogen defense overrides UV protection through an inversely regulated ACE/ACE type of light-responsive gene promoter unit. Proc Natl Acad Sci USA 99:2428–2432

Manning WJ, von Tiedemann A (1995) Climate change: potential effects of increased atmospheric carbon dioxide (CO_2), ozone (O_3), and ultraviolet-B (UV-B) radiation on plant diseases. Environ Pollut 88:219–245

Meagher RB (2000) Phytoremediation of toxic elemental and organic pollutants. Curr Opin Plant Biol 3:153–162

Mithöfer A (2002) Suppression of plant defence in rhizobia-legume symbiosis. Trends Plant Sci 7:440–444

Moesta P, Grisebach H (1980) Effects of biotic and abiotic elicitors on phytoalexin metabolism in soybean. Nature 286:710–711

Neill S, Desikan R, Hancock J (2002) Hydrogen peroxide signaling. Curr Opin Plant Biol 5:388–395

Nover L, Neumann D, Scharf K-D (1989) Heat-shock and other stress responses of plants. Springer, Berlin Heidelberg New York

Nürnberger T, Scheel D (2001) Signal transmission in the plant immune response. Trends Plant Sci 6:372–379

Pell EJ, Schlagnhaufer CD, Arteca RN (1997) Ozone-induced oxidative stress: mechanisms of action and reaction. Physiol Plant 100:264–273

Powles SB, Holtum JAM (eds) (1994) Herbicide resistance in plants. Biology and biochemistry. Lewis Publishers, Boca Raton

Quirino BF, Noh Y-S, Himelblau E, Amasino RM (2000) Molecular aspects of leaf senescence. Trends Plant Sci 5:278–282

Ramade F (ed) (1979) Ecotoxicology. Wiley, Chichester, UK

Rao MV, Davis KR (2001) The physiology of ozone induced cell death. Planta 213:682–690

Reymond P, Farmer EE (1998) Jasmonate and salicylate as global signals for defense gene expression. Curr Opin Plant Biol 1:404–411

Ries G, Heller W, Puchta H, Sandermann H Jr, Seidlitz HK, Hohn B (2000) Elevated UV-B radiation reduces genome stability in plants. Nature 406:98–101

Sandermann H (1992) Plant metabolism of xenobiotics. Trends Biochem Sci 17:82–84

Sandermann H (1994) Higher plant metabolism of xenobiotics: the 'green liver' concept. Pharmacogenetics 4:225–241

Sandermann H (1996) Ozone and plant health. Annu Rev Phytopathol 34:347–366

Sandermann H Jr (2000) Active oxygen species as mediators of plant immunity: three care studies. Biol Chem 381:649–653

Sandermann H Jr, Ernst D, Heller W, Langebartels C (1998) Ozone: an abiotic elicitor of plant defence reactions. Trends Plant Sci 3:47–50

Schäffner A, Messner B, Langebartels C, Sandermann H (2002) Genes and enzymes for *in-planta* phytoremediation of air, water and soil. Acta Biotechnol 22:141–152

Scheel D, Wasternack C (eds) (2002) Plant signal transduction. Oxford University Press, Oxford

Scheideler M, Schlaich NL, Fellenberg K, Beissbarth T, Hauser NC, Vingron M, Slusarenko A, Hoheisel JD (2002) Monitoring the switch from housekeeping to pathogen defense metabolism in *Arabidopsis thaliana* using cDNA arrays. J Biol Chem 277:10555–10561

Schenk PM, Kazan K, Wilson I, Anderson JP, Richmond T, Somerville SC, Manners JM (2000) Coordinated plant defense responses in *Arabidopsis* revealed by microarray analysis. Proc Natl Acad Sci USA 97:11655–11660

The Arabidopsis Genome Initiative (2000) Analysis of the genome sequence of the flowering plant *Arabidopsis thaliana*. Nature 408:796–812

Vogt T, Jones P (2000) Glycosyltransferases in plant natural product synthesis: characterization of a supergene family. Trends Plant Sci 5:380–386

Vranová E, Atichartpongkul S, Villarroel R, Van Montagu M, Inzé D, Van Camp W (2002) Comprehensive analysis of gene expression in *Nicotiana tabacum* leaves acclimated to oxidative stress. Proc Natl Acad Sci USA 99:10870–10875

Walling LL (2000) The myriad plant responses to herbivores. J Plant Growth Regul 19:195–218

Walter MH (1989) The induction of phenylpropanoid biosynthetic enzymes by ultraviolet light or fungal elicitor in cultured parsley cells is overriden by a heat-shock treatment. Planta 177:1–8

Werck-Reichhart D, Hehn A, Didierjean L (2000) Cytochromes P450 for engineering herbicide tolerance. Trends Plant Sci 5:116–123

Wittstock U, Gershenzon J (2002) Constitutive plant toxins and their role in defense against herbivores and pathogens. Curr Opin Plant Biol 5:300–307

Xiong L, Zhu J-K (2001) Abiotic stress signal transduction in plants: molecular and genetic perspectives. Physiol Plant 112:152–166

Zasloff M (2002) Antimicrobial peptides of multicellular organisms. Nature 415:389–395

2 Metabolism of Natural and Xenobiotic Substrates by the Plant Glutathione S-Transferase Superfamily

ROBERT EDWARDS and DAVID P. DIXON

2.1 Plant Glutathione Transferase (GST) Superfamily

2.1.1 Discovery

Glutathione transferases, also referred to as glutathione S-transferases (GSTs; EC 2. 5.1.18), were first described in animals in the 1960s (Booth et al. 1961), where they were determined to have important roles in catalysing the conjugation of drugs with the tripeptide glutathione (γ-glutamyl-cysteinyl-glycine). In this detoxification reaction the cysteinyl SH-group of the glutathione serves as a nucleophile, which attacks an electrophilic centre in the drug to give rise to the S-glutathionylated derivative via substitution, or, more rarely, addition reactions (Fig. 2.1). The discovery of plant GSTs soon followed that of their mammalian counterparts. The identification of a glutathione-conjugated metabolite of atrazine in maize (see Fig. 2.1A; Shimabukuro and Swanson 1969) quickly led to the partial purification and characterisation of a maize GST that catalysed this reaction (Frear and Swanson 1970). These early studies clearly identified GSTs as a primary determinant of atrazine selectivity, with cereal crops containing much greater herbicide-detoxifying GST activity than competing weeds. Interests in the role of GSTs in herbicide selectivity in crops and weeds then became a major driver for further work on these enzymes for much of the next 30 years. GSTs involved in herbicide metabolism have since been purified and then cloned from major crops (maize, soybean, wheat and rice) as well as problem weeds (reviewed by Edwards and Dixon 2000). It has also been recognised that GST activities toward xenobiotics appear to be ubiquitous in both lower and higher plants (Pflugmacher et al. 2000) as well as in algae (Tang et al. 1998). Although most work has concentrated on the involvement of GSTs in herbicide selectivity, GSTs active in conjugating model substrates such as 1-chloro-2,4-dinitrobenzene (CDNB) without specific roles in herbicide detoxification in the field have been purified to homogeneity from a diversity of plants including sugar

Ecological Studies, Vol. 170
H. Sandermann (Ed.)
Molecular Ecotoxicology of Plants
© Springer-Verlag Berlin Heidelberg 2004

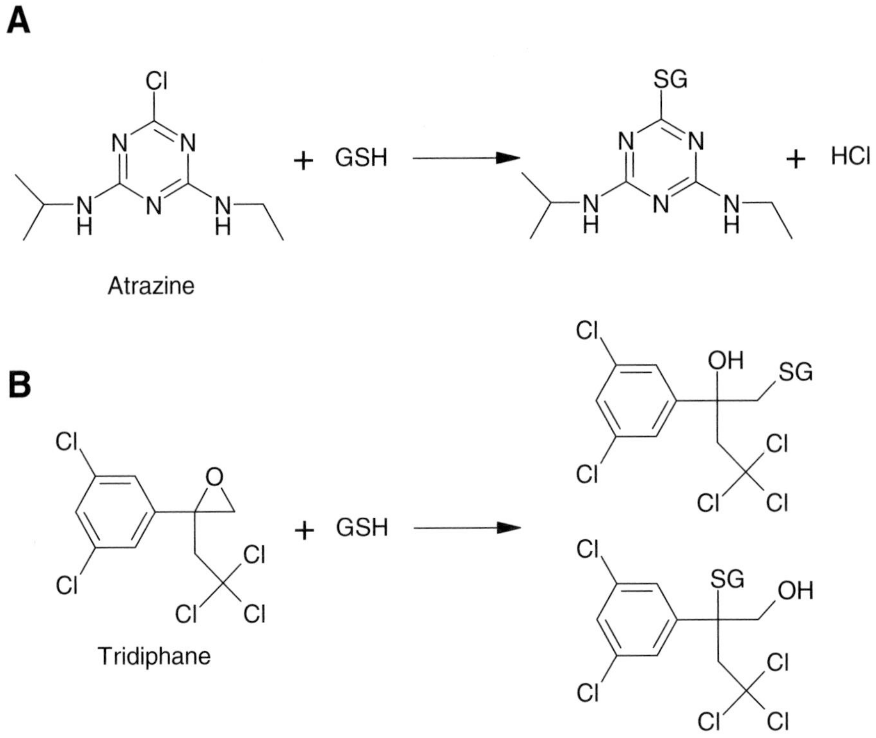

A

Atrazine

B

Tridiphane

Fig. 2.1. Detoxification of xenobiotics by GSTs mediated by catalysing their conjugation with glutathione by substitution (A) or addition (B) reactions. **A** shows the glutathione conjugation of the chloro-s-triazine herbicide atrazine, while **B** shows the two possible conjugation reactions seen with the atrazine synergist tridiphane. (Lamoureux and Russness 1986)

cane (Singhal et al. 1991), Norway spruce (Schröder and Berkau 1993), pumpkin (Fujita et al. 1995), broccoli (Lopez et al. 1994) and peas (Edwards 1996). In all these studies, the importance of these enzymes in xenobiotic detoxification has been to the fore, and this preoccupation with the roles of GSTs in foreign compound metabolism has detracted from our understanding of the roles of these enzymes in both constitutive and stress-invoked plant physiology and metabolism. Significantly, our awareness of the importance of GSTs in endogenous metabolism in plants has been largely fuelled by the observations that transcripts encoding GSTs accumulate substantially during cell division, or in response to biotic stress, invoked by changes in the environment or by infection (reviewed by Marrs 1996; Edwards et al. 2000). These GST transcripts have typically been identified using subtractive or differential screening of cDNAs, resulting in a large number of GST sequences entering the public databases. In addition, this genetic information has recently been greatly extended with the availability of genome and comprehensive expressed sequence tagged (EST) sequence libraries. With the availability of this genome information has come the identification of a surprisingly large and diverse family of GSTs and related enzymes in plants (Dixon et al. 2002b).

2.1.2 Organisation and Nomenclature of GST Gene Superfamily

The unexpected size and complexity of the plant GST superfamily has necessitated a re-evaluation of the best way to classify these enzymes. Until recently, almost all characterised GSTs were mammalian GSTs, which were classified into the alpha, mu, pi and theta classes. The theta class was a very heterogeneous class and contained most of the non-mammalian enzymes, including all the plant GSTs (Wilce and Parker 1994). From the wealth of GST sequence data that became available for plants and other organisms, it was clear that this classification was rapidly becoming untenable. Subsequently, a number of attempts were made to reclassify plant GSTs, each recognising more classes than the previous system (Droog et al. 1995; Droog 1997; Dixon et al. 1998; Edwards et al. 2000). The current system (Dixon et al. 2002b) identifies five distinct classes, each given a Greek letter: phi, tau, theta, zeta and lambda. In addition, a sixth class comprising the glutathione-dependent dehydroascorbate reductases (DHARs) has also been identified as being GST-derived. One shortcoming of this classification scheme is that by identifying GSTs solely on the basis of their sequence similarity to existing enzymes, many genes that are members of the superfamily may not encode proteins that show any functional activity as glutathione-conjugating enzymes. Instead, as we shall see, these proteins have evolved to use glutathione for a variety of alternative catalytic and non-catalytic functions. However, in terms of conservation in structure, these diverse proteins remain closely related to the "true" GSTs and can be most usefully considered members of the GST superfamily. Of the family members, the small zeta and theta classes have the closest relatives in mammals and other organisms. The remaining four classes appear to be plant-specific, with the large phi and tau GSTs associated with the classic detoxifying activities toward herbicides.

In order to use this classification system with minimal confusion, a universal nomenclature has been developed based on the system used for mammalian GSTs (Dixon et al. 2002b). This system identifies both the originating species and the class into which the GST falls. Firstly, each GST class has been assigned a letter, with F denoting the phi class, U = tau, T = theta, Z = zeta and L = lambda. The GST is then prefixed with the species name. Thus, *At*GSTF would denote a phi class GST from *Arabidopsis thaliana*. Finally, each GST subunit is assigned a number, based on the clustering of the genes from that GST class where genome organisation is known, or if not known, on the order of discovery (Wagner et al. 2002). Since GSTs are dimeric proteins composed of either identical or distinct subunits derived from the same class, this system can uniquely identify all potential GST enzymes. For example, AtGSTF2–2 is a homodimer from *Arabidopsis* of phi number 2 subunits, while AtGSTF2–3 is the corresponding heterodimer composed of phi subunits 2 and 3. The respective genes are then named similarly, but in lower case (e.g., gstf2).

In addition to the soluble GSTs described above, glutathione conjugation reactions can also be catalysed by unrelated proteins. In animals an unrelated family of microsomal, trimeric enzymes described as membrane-associated proteins of eicosanoid and glutathione metabolism (MAPEG), which possess limited glutathione-conjugating activities, have been described (Jakobsson et al. 1999). A screen of the EST and genome databases reveals that similar genes are present in plants, though the functions of the respective proteins have not been studied. Glutathione conjugation reactions can also be effected by proteins that do not bind directly with the tripeptide. In maize, the enzyme responsible for conjugating cinnamic acid to glutathione was found to be an ascorbate peroxidase, and probably catalysed the reaction by generating thiyl radicals of glutathione, which then drove an addition reaction with the cinnamic acid (Dean and Devarenne 1997). A somewhat analogous reaction has also been reported to be mediated by soybean lipoxygenase, which in the course of acting on linoleic acid can oxidise glutathione to give the respective thiyl radical, which reacts readily to conjugate ethacrynic acid (Kulkarni and Sajan 1997).

2.1.3 Common Structural and Functional Characteristics of Plant GSTs

Since membership of the GST superfamily is based on sequence similarity, all its members share certain characteristics and are all assumed to be derived from a single ancestral protein. The soluble GSTFs, GSTTs, GSTUs and GSTZs are all polypeptides of around 25 kDa, which in each case associate with other subunits of the same class to form dimers (Dixon et al. 1999; Sommer and Böger 1999). In plants, these dimers can be either homodimers or heterodimers, and this mixing and matching of subunits within each class can give rise to considerable isoenzyme multiplicity. Crystal structures exist for many of the mammalian GST classes (Dirr et al. 1994; Armstrong 1997; Sheehan et al. 2001). Recently, the structures of phi class GSTs from maize (Neuefeind et al. 1997a, b) and *Arabidopsis* (Reinemer et al. 1996), a zeta class GST from *Arabidopsis* (Thom et al. 2001) and a tau class GST from wheat (Thom et al. 2002) have been reported. For illustration, the structure of the dominant GSTF from maize binding the glutathione conjugate of the herbicide atrazine is shown in Fig. 2.2. All these crystallised enzymes share a remarkably conserved overall structure, despite the limited similarity in polypeptide sequences. All GSTs contain two structurally distinct domains, separated by a linker region. The smaller N-terminal domain contains the residues that

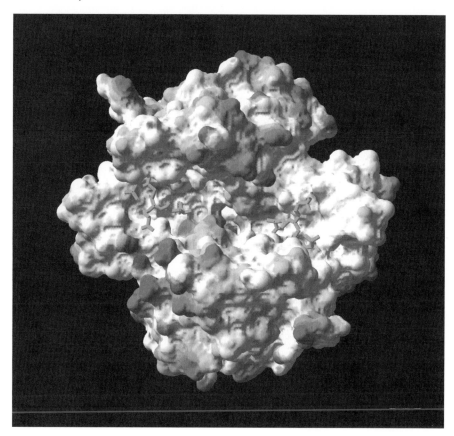

Fig. 2.2. (in color at end of book) The three-dimensional structure of *Zm*GSTF1–1, the major phi class GST expressed in maize foliage. The two *Zm*GSTF1 polypeptides are shown dimerised together, with the glutathione conjugate of the herbicide atrazine (shown in *green*) at the respective active sites. The structure represents the projected molecular surface and is coloured to represent the calculated electrostatic potential (*red* negative; *blue* positive)

interact with glutathione bound at the active site, and together form the G-site. Since GSTs are highly specific towards GSH, or a close analogue such as homoglutathione (γ-glutamyl-cysteinyl-β-alanine), this region of the polypeptide shows much less structural diversity than the rest of the protein. The larger C-terminal domain includes many of the residues responsible for binding the co-substrate; these form the hydrophobic site (H-site). Because of the wide variation in substrate specificity shown between different GSTs, it is not surprising that the C-terminal domain differs substantially between different classes.

The significance of the dimerisation of GST subunits to form functional GSTs is not understood. The dimerisation interface varies substantially between classes, ensuring dimers typically only form between GST monomers of the same class, and each monomer is believed to have a self-contained active site located in the cleft formed between the subunits with little or no interaction with the adjacent polypeptide. For GSTs derived from subunits of the same class, these active sites appear catalytically as well as physically independent, as shown by various enzyme studies (Danielson and Mannervik 1985; Dixon et al. 1999). However, it has recently been demonstrated that mammalian pi and mu subunits can form mixed class dimers under certain conditions, with the resulting heterodimer showing that the subunit interactions can affect activity (Pettigrew and Colman 2001). Although conventional GSTs are invariably dimers, we have recently identified outlying members of the GST family, such as the DHARs and lambda GSTs, which are monomers (Dixon et al. 2002a).

GSTs catalyse their reactions by promoting the formation of the reactive thiolate anion (GS–) of glutathione, effectively lowering the pKa of glutathione to values similar to physiological pH. This is achieved by stabilising the thiolate anion through hydrogen bonding with an adjacent hydroxy group, provided by a tyrosine residue in most mammalian GSTs and a serine residue in most plant GSTs. Lambda GSTs and DHARs have cysteine residues at the corresponding location; this has major implications for their activity as instead of stabilising the thiolate anion of the tripeptide, they form mixed disulphides with glutathione. These enzymes cannot therefore catalyse conjugation reactions, but at least in the case of DHARs are instead involved in redox reactions (Dixon et al. 2002a).

2.1.4 GSTs of *Arabidopsis thaliana*: a Blueprint for Studying Plant GSTs

The availability of the complete genome of the model plant *Arabidopsis thaliana* has made it possible for the first time to examine the entire complement of GSTs and related proteins making up the GST superfamily in a plant. In total, 52 members of the GST superfamily have so far been identified in the genome (Dixon et al. 2002b; Wagner et al. 2002). Most of these sequences have also been identified in the *Arabidopsis* expressed sequence tag (EST) databases, demonstrating that the majority of the GST genes are expressed and are not pseudogenes. Of these 52, over half (28) are tau class, and most of the rest (13 of 24) are phi class. The remainder are three theta GSTs, two zeta GSTs, two lambda GSTs and four DHARs. Additionally, *Arabidopsis* contains a single unrelated microsomal GST gene. Many of the *Arabidopsis* GSTs are present in clusters of up to seven similar genes, strongly indicative of gene duplication events, although the reasons for such extensive duplication are unknown. Interestingly, the sequences for the majority of the GST genes predict that the respective proteins will be expressed in the cytoplasm. The exception is seen with the DHARs and GSTLs, where from the N-terminal leader sequences it can be predicted that one protein from each class will be expressed in the

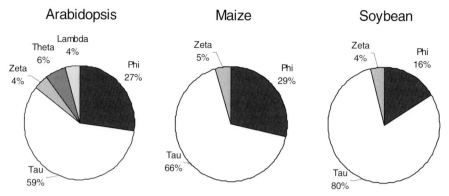

Fig. 2.3. Pi charts showing the relative abundance of each class of GST genes in *Arabidopsis*, maize and soybean. For *Arabidopsis*, the proportions of GSTs were determined from the genome sequence (Anonymous 2001). For maize and soybean, the numbers of GSTs were determined by the analysis of EST databases and as such do not necessarily depict the full range of GST genes present. (McGonigle et al. 2000)

chloroplast. Additionally, one phi class GST has a chloroplast targeting signal (Wagner et al. 2002).

In *Arabidopsis*, the dominance of tau GST genes in the genome and in the EST databases suggests that they are the most abundant GSTs in this model plant (Fig. 2.3). Recent genomic studies in maize and soybean have also shown that in terms of numbers gstu genes are more abundant than gstf genes, with maize having 12 gstfs and 28 gstus, while soybean has 4 gstfs and 20 gstus (Fig. 2.3). However, when the total number of ESTs for each class was compared as a measure of the abundance of the respective mRNAs, the ratio of phi to tau transcripts was 57:43 % in maize, while in soybean the ratio was 94:6 % (McGonigle et al. 2000). Significantly, the relative abundance of tau and phi mRNAs closely mirrors what is seen at the biochemical level, with GSTFs being the dominant class in maize (Jepson et al. 1994), while in soybean GSTUs are the major xenobiotic-detoxifying GSTs (Flury et al. 1995; Andrews et al. 1997a; Skipsey et al. 1997).

The usefulness of *Arabidopsis* in GST research is not limited to its genome sequence. A large population of tagged T-DNA knockout lines exists, and identification of GST knockouts and subsequent analysis of these lines should prove useful in determining GST function, particularly for enzymes in classes with few members where duplication of function is unlikely to be a problem. Even where gene duplication masks obvious phenotypes, the increasing availability of DNA arrays and proteomics will provide powerful tools to dissect the natural functions of GSTs and help define their roles in stress tolerance and xenobiotic detoxification.

2.2 Glutathione Transferases and Stress Responses

2.2.1 Roles for GSTs in Stress Tolerance in Animals

Because of their long association with detoxification and their inducibility by
chemicals and changes in environment, there is good precedence to believe
that GSTs are important proteins in counteracting cellular stress. Indeed, several authors have proposed that GSTs play a central role in the glutathione-
dependent antioxidant defence system of animal cells (Hayes and Mclellan
1999; Cnubben et al. 2001). Before considering the roles for GSTs in stress tolerance in plants, it is therefore well worth considering what we know about
the roles of these proteins in stress tolerance in animals. Interestingly, while
the GSTTs and GSTZs show conservation in sequence and function between
plants and animals (see Sects. 2.3.1.2 and 2.3.3.1), all other classes of GSTs
have arisen independently. Also, while the conserved GSTs appear to be constitutively expressed and fulfil essential housekeeping functions in the
eukaryotes studied (Fernández-Cañón et al. 1999), the other classes, which
typically have the greatest activities in xenobiotic detoxification, are frequently stress inducible. It is therefore worthwhile considering whether or not
plant-specific and animal-specific GSTs have independently evolved to fulfil
similar, or different, roles in stress tolerance.

The alpha, mu, pi and sigma classes are classically associated with the
detoxification of drugs and pollutants, notably in the liver, with the resulting
glutathione conjugates being processed to mercapturic acid conjugates prior
to excretion (Habig et al. 1974). The importance of these enzymes in protecting against the toxic effects of drugs has recently been demonstrated using
gene knockouts in mice (Ruscoe et al. 2001). While GSTP knockout mice
appeared normal, they were highly sensitive to the carcinogenic effects of
polycyclic aromatics when applied to the skin, as well as to the cytotoxic
effects of cis-platinum. Unexpectedly, GSTP knockout mice were more resistant to the hepatotoxic effects of acetaminophen, a drug normally associated
with detoxification by glutathione conjugation, and it was demonstrated that
the GSTP was playing an unexpected role in promoting toxicity, which was
not related to the rates of acetaminophen detoxification (Henderson et al.
2000). Such studies illustrate how little we really understand these enzymes,
even when they are playing a "traditional" role as in drug metabolism. Similarly, in some instances, glutathione conjugation mediated by GSTs actually
increases the toxicity of the xenobiotic. For example, hydroquinone, bro-
mobenzene and dihaloalkanes are all conjugated by GSTs to produce reactive
mutagenic glutathione derivatives (Hinson and Forkert 1995; Sherratt et al.
1998).

The majority of studies examining the functions of GSTs in animals have
concentrated on their roles in drug detoxification rather than in the metabo-

lism of naturally occurring stress metabolites. Recent in vitro studies with different classes of mammalian GSTs have shown that these enzymes are able to catalyse the conjugation of toxic alkenal and epoxide derivatives of polyunsaturated fatty acids and cholesterol α-oxide all arising from membrane oxidation (reviewed by Hayes and McLellan 1999). Analogous detoxification reactions have also been reported with adenine and thymine propenals derived from oxidative damage of DNA and the quinone metabolites formed from the oxidation of catecholamines and oestrogens (Hayes and Mclellan 1999). Examples of conjugation reactions with endogenous metabolites are shown in Fig. 2.4. Significantly, the importance of these detoxification reactions in vivo is supported by the observation that a mercapturic acid conjugate of the major stress metabolite 4-hydroxynon-2-enal has been identified in rat urine (Alary et al. 1995).

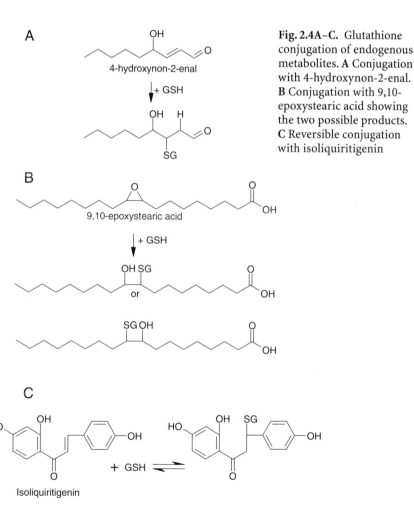

Fig. 2.4A–C. Glutathione conjugation of endogenous metabolites. **A** Conjugation with 4-hydroxynon-2-enal. **B** Conjugation with 9,10-epoxystearic acid showing the two possible products. **C** Reversible conjugation with isoliquiritigenin

As well as acting on the degradation products derived from oxidation, GSTs can also demonstrate a direct protective activity against the primary products of membrane oxidation by acting as glutathione peroxidases (GPOXs) toward organic hydroperoxide substrates. GPOXs catalyse the reaction:

$$ROOH + 2\,GSH \rightarrow ROH + H_2O + GSSG$$

where R can be H (for hydrogen peroxide) or an organic molecule. The oxidised glutathione (GSSG) formed is then re-reduced by NADPH-dependent glutathione reductases. This reaction is extremely important in preventing cytotoxicity as the accumulation of hydroperoxides leads to the formation of highly reactive peroxyradicals and alkenal degradation products (Hayes and Mclellan 1999). The reduction of hydroperoxides to the respective alcohols is catalysed by both GSTs and a group of unrelated selenium-dependent GPOXs (Wendel 1981; Eshdat et al. 1997). These two groups of GPOXs have a differing spectrum of activities, with the GST-type GPOXs only having activity with organic hydroperoxides, while the selenium-dependent enzymes will also reduce hydrogen peroxide. Typical substrates of GST-GPOXs include the hydroperoxides of lipids and sterols, and there is good evidence that GST-GPOXs play an important role in preventing the accumulation of such compounds in mammalian cells exposed to oxidative stress (Yin et al. 2000).

From this brief review, we can therefore see that GSTs form an important part of the defence system of mammalian cells directed at counteracting oxidative stress. Significantly, with the exception of the theta GSTs, plants have had to evolve similar protective activities independently of their mammalian counterparts.

2.2.2 Regulation of GSTs in Plants by Biotic and Abiotic Stress

GSTs have now become virtually characteristic markers of stress induction in plants, particularly at the level of their gene induction. Several early studies established that GST activities in cereal crops were enhanced following exposure to sub-lethal doses of herbicides and that this enhancement could be linked to a subsequent increased herbicide tolerance (Dean et al. 1990; Alla 1995; Uotila et al. 1995; Rossini et al. 1998). More significantly, it was determined that herbicide safeners could also selectively increase GST activities toward pesticides in cereals, thereby increasing herbicide tolerance in crops (see Sect. 2.4.2). Evidence that GSTs were also enhanced in plants exposed to biotic stress subsequently emerged in a number of studies that determined increases in their enzyme activity following a range of treatments (reviewed by Marrs 1996; Edwards et al. 2000). These included drought stress, infection by microbial pathogens, exposure to UV light, ethylene or auxins and treatment with heavy metals and xenobiotics, which are neither herbicides nor safeners. In every case where correlations between enhancement of GST activity and expression of the respective genes have been examined, it appears that increases in GST activity can be accounted for by an increase in the levels of the respective proteins brought about by increased translation of the respective mRNA resulting from gene induction (Jepson et al. 1994). However, in many cases while it has been reported that transcripts apparently encoding GSTs accumulate in response to these diverse stress treatments, the subsequent effect on GST activities *in planta* has not been reported.

In view of the large number of GSTs that have been reported to be induced in different plants, there have been few attempts to co-ordinately study the differential regulation of GSTs that make up this superfamily. In wheat, it was reported that the expression of the phi GSTA1 protein was enhanced following an infection with *Erisyphe graminis*, or treatment with fungal elicitors, but not by exposure to herbicides or other chemical treatments (Mauch and Dudler 1993). Conversely, two tau class GSTs were only induced by the chemical treatments. Recent studies in *Arabidopsis* have extended the analysis of the differential expression of multiple GST genes, by infection, chemicals that perturb redox homeostasis and xenobiotics (Dixon et al. 2002b; Wagner et al. 2002). One of the interesting observations derived from these studies has been the inability to predict the regulation of GSTs based on their classification. Thus, GSTs within a class may be quite distinctly regulated by different stimuli, while GSTs across classes can show co-regulation. Co-expression of GSTs of different classes suggests they have identical, or directly complementary, functions. Such an hypothesis is supported by biochemical studies carried out on herbicide detoxifying activity (Cummins et al. 1997a) and flavonoid transport (Alfenito et al. 1998), which show that selected GSTFs and GSTUs can complement one another's functional roles even though their sequences show little identity.

Although the functional significance of GST induction is still largely unknown, the signalling mechanisms regulating the expression of the respective genes has been studied in some detail. In animals, GSTs and other inducible drug detoxifying enzymes contain conserved *cis* elements in the promoters of the respective genes termed xenobiotic responsive elements (XREs) and antioxidant response elements (AREs) respectively (Hayes and Pulford 1995). As their names suggest these *cis* elements bind to regulatory proteins that promote gene transcription when the cell is exposed to xenobiotics (XRE) or oxidative stress (ARE). For example, the alpha and pi GSTs have AREs in their respective promoters, which are activated by binding the transcription factor Nrf2, which is selectively released during oxidative stress (Hayes and McMahon 2001). A thorough overview of *cis*-elements and transcription factors regulating plant gene expression is provided in Chapter 6. Initially, it was thought that the promoters of *gst* genes involved in herbicide metabolism would contain similar regulatory elements, but to date no ARE or XRE sequences have been identified in plant *gst* promoters. Instead, it has been necessary to identify plant-specific signalling elements in *gst* genes

(Marrs 1996). One promoter element identified in stress-inducible plant *gsts* is the 20-bp octopine synthase (*ocs*) element, also termed the activating sequence 1 (*as-1*) element (Xiang et al. 1996). Promoters containing the *as-1* *cis*-acting element confer inducibility of GSTUs to auxins, salicylic acid and methyl-jasmonate in tobacco, with *as-1* showing some similar characteristics to the ARE elements (Droog et al. 1995; Ulmasov et al. 1995; Xiang et al. 1996). However, the spectrum of other chemicals that activate such promoters appears to be dependent on the number and organisation of *as-1* elements. Thus, an *as-1*-containing promoter of a GSTU from soybean was induced by a wider range of compounds, including heavy metals (Ulmasov et al. 1995), than was the case with an *as-1*-containing promoter of a GSTU from tobacco (Xiang et al. 1996). Interestingly, the promoters of tobacco GSTUs also contain multiple specific auxin regulatory elements, which regulate their responsiveness to auxins (Droog et al. 1995; Takahashi et al. 1995). In the case of the GSTZs in carnation, a further distinct regulatory element was identified that resulted in responsiveness to ethylene. These ethylene responsive elements (EREs) were responsible for enhancing the expression of two GSTZs in the course of petal senescence (Itzhaki and Woodson 1993). Similarly, other promoter elements have been described that are regulated by either infection by fungi or are co-regulated with genes involved in anthocyanin biosynthesis (Marrs 1996).

In summary, from what is known of the gene regulation of GSTs, it would be anticipated that a number of parallel as well as convergent signalling pathways operate during stress responses. Some of these signalling pathways appear to be initiated by single well-defined agents such as ethylene and auxins, which act on the ERE and auxin-responsive promoter elements, respectively. Other stimuli, which may include these hormones but also include xenobiotics and the products of oxidative stress, may be independently recognised by receptor systems, which then convergently activate more 'general purpose' promoter elements such as *as-1*. While this model of GST regulation is likely to be an over-simplification of what is seen *in planta*, it is a useful starting point to consider how this large gene family can be selectively responsive to such diverse stimuli through a number of specific regulation elements in the promoters of individual *gst* genes.

2.3 Roles for Plant GSTs in Endogenous Metabolism

2.3.1 Activities of GSTs in Detoxifying Metabolites of Oxidative Stress

It is very likely that the known roles of GSTs in detoxifying products of lipid and DNA oxidation in animals would be even more essential in plants, due to the constant production of reactive oxygen species (ROS), such as hydroperoxide (H_2O_2) and superoxide (O_2-), hydroxyl ($OH-$) and hydroperoxyl (H_2O-) radicals, as by-products of photosynthesis. Plants contain an efficient and well-coordinated antioxidant system that uses ascorbate, glutathione, tocopherol and secondary metabolites to mop up these ROS (Noctor and Foyer 1998). However, under adverse conditions such as extreme temperatures or irradiance, or in the course of stress imposed by drought, chemicals or infection, it is inevitable that the production of ROS exceeds the capacity of the endogenous antioxidant system. ROS accumulation in plants is known to rapidly lead to the oxidation of membranes, resulting in the formation of lipid hydroperoxides, such as linoleic acid hydroperoxide, which subsequently degrade to evolve cytotoxic alkenals such as malondialdehyde (Noctor and Foyer 1998). As determined in animals (Hayes and Mclellan 1999), GSTs can detoxify these oxidatively modified metabolites by both catalysing the reduction of the hydroperoxides and the glutathione conjugation of the electrophilic degradation products. The evidence for plant GSTs functioning with these two activities will be considered in turn.

2.3.1.1 Conjugation of Oxidation Products

Unfortunately, unequivocal evidence that GSTs conjugate lipid or DNA oxidation products in oxidatively stressed plants has yet to be demonstrated, as there are no reports of the respective glutathione conjugates accumulating *in planta*. However, GSTFs active in conjugating 4-hydroxynonenal have been identified in sorghum (Gronwald and Plaisance 1998), and an analogous glutathione addition reaction has been determined with the substrates ethacrynic acid or crotonaldehyde with both GSTU and GSTF isoenzymes in peas (Edwards 1996), wheat (Cummins et al. 1997a), maize and soybean (McGonigle et al. 2000). It would be anticipated that following their formation that transport glutathione conjugates of lipid degradation products would be rapidly imported into the vacuole by ABC transporters, which show high activities toward such conjugates (Rea et al. 1998).

2.3.1.2 Activity as Glutathione Peroxidases

Plant GSTs showing GPOX activity have been identified from the phi, tau and theta classes, but not from the zeta, lambda or DHAR classes (Dixon et al. 2002a; Thom et al. 2002). Of all the plant GSTs tested to date, the greatest GPOX activity has been found to be associated with the constitutively expressed GSTT class (Dixon et al. 2001). GPOX activity has also been demonstrated with both synthetic organic hydroperoxides, such as cumene hydroperoxide, and naturally occurring hydroperoxides of linoleic acid, with both phi and tau GSTs from *Arabidopsis* (Bartling et al. 1993), sorghum (Gronwald and Plaisance 1998) and wheat (Cummins et al. 1997a). Clues to the importance of these enzymes in stress tolerance in plants have been obtained from studying their regulation and from transgenic studies. The GST-GPOX from *Arabidopsis* was constitutively expressed during normal plant growth and following treatment with plant hormones, but its expression varied with the transition to bolting and flowering, suggesting the enzyme had a role in plant development (Bartling et al. 1993). Good evidence for a role for GST-GPOXs in stress tolerance was obtained when a GSTF showing GPOX activity toward cumene hydroperoxide was over-expressed in transgenic tobacco (Roxas et al. 1997). The resulting seedlings showed enhanced growth under chilling or salt stress and accumulated oxidised glutathione, a reaction product associated with the reduction of organic hydroperoxides. It was hypothesised that rather than providing antioxidant stress protection directly, the GPOX mediated an increase in cellular GSSG, which then acted as an inducing signal for other protective enzymes. GSTs with GPOX activity have also been associated with preventing oxidative cell death, which occurs during apoptosis. When expressed in yeast, a GSTU from tomato with limited GPOX activity prevented the damage associated with the expression of the Bax protein, which induces oxidative cell death as well as providing protection against exogenous treatments of hydrogen peroxide (Kampranis et al. 2000).

The GPOX activity of GSTs has also been implicated in herbicide tolerance in grass weeds. Populations of black-grass (*Alopecurus myosuroides*) showing resistance to multiple classes of herbicides all contained unusually high levels of a GSTF, which was very active as a GPOX (Cummins et al. 1999). It was demonstrated that these plants accumulated reduced levels of organic hydroperoxides when exposed to different herbicides, and it was proposed that this GST-GPOX exerted a broad-ranging protective effect against cytotoxic hydroperoxides, which accumulated as a result of herbicide-invoked derailed metabolism. Interestingly, treatment of maize and wheat with herbicide safeners results in the accumulation of GSTUs and GSTFs, which differ from the respective constitutive GSTs in showing high GPOX activity (Cummins et al. 1997a; Cole et al. 1998). It is therefore tempting to speculate that these safener inducible GSTs exert a protective effect at two levels: firstly, through enhancing the glutathione conjugation and detoxification of herbicides and secondly through reducing the amount of oxidative injury they invoke.

2.3.2 GSTs and Plant Secondary Metabolism

2.3.2.1 Conjugation of Secondary Metabolites

Because of their well-known role in detoxifying xenobiotics, it has been a natural progression to look for evidence that these enzymes conjugate plant natural products (Kolm et al. 1995). In animals, a protective function in conjugating plant toxins is often cited as a primary reason for the evolution of GSTs in herbivores (Sheehan et al. 2001). Surprisingly, there is little chemical evidence that glutathione conjugation reactions are commonplace *in planta* (reviewed by Lamoureux and Rusness 1993). Thus, a glutathione conjugate of caftaric acid and a sulphur-containing gibberellin derivative are among the few natural products identified that could have arisen from S-glutathionylation. The absence of glutathionylated natural products, or their sulphur-containing derivatives, is surprising given the abundance of GSTs and glutathione in plant cells. In addition plant cells contain an efficient transporter system for glutathione conjugates in the vacuole membrane (Rea et al. 1998) and a set of hydrolytic enzymes that can rapidly process glutathione conjugates in the vacuole (Wolf et al. 1996). Thus, all the machinery for making, transporting and processing glutathione conjugates is present, yet the system can only be seen to operate with xenobiotic substrates. One rational explanation for this conundrum is that the natural products subject to this pathway are either very rapidly processed to polar derivatives that have not yet been described, or that the conjugates formed are unstable (Edwards et al. 2000). Significantly, researchers have been able to find glutathione conjugates of natural products when they have been selectively sought. When chickpea cell cultures were fed with radiolabelled cinnamic acid, the respective glutathione conjugate transiently accumulated (Barz and Mackenbrock 1994). Although this conjugate arises from the action of an ascorbate peroxidase rather than a GST (Dean and Devarenne 1997), the formation of S-cinnamoyl-glutathione demonstrated that conjugation of toxic natural products can be observed *in planta* with the respective conjugate serving as an excellent ligand for active uptake by the vacuole (Walczak and Dean 2000).

The evidence for other endogenous natural products undergoing glutathione conjugation is largely based on in vitro studies. We have determined that 4,2',4'-trihydroxychalcone (isoliquiritigenin) can undergo addition reactions with glutathione (Fig. 2.4). However, this reaction was found to proceed spontaneously at physiological pH and was not catalysed by any of the GSTFs or GSTUs tested (Cummins et al. 2003). Significantly, the conjugates formed were found to be very sensitive to the conditions used in extraction and analysis, being unstable at extreme pH. Quinone methides of the flavone quercetin are also reported to undergo reversible *S*-glutathionylation (Boersma et al. 2000), and it is possible that such reversible conjugation may be more important than previously recognised in detoxifying reactive quinone derivatives of phenolics formed during oxidative events like the hypersensitive response, when oxidised phenolics are known to accumulate (Somssich and Hahlbrock 1998). However, it is debatable whether or not these conjugation reactions require the intervention of GSTs, as the flavonoid glutathione derivatives are particularly potent competitive inhibitors of these enzymes (Mueller et al. 2000).

Interest in glutathione conjugation of natural products has tended to concentrate on endogenous metabolites. It is also worth bearing in mind that plants have to metabolise natural product toxins generated by invading microorganisms and by competing allelopathic plants. A good example of such an allelochemical are the isothiocyanates, reactive alkylating agents released from their respective glucosinolate precursors and used as allelopathic chemicals by cruciferous plants (Bartling et al. 1993). These isothiocyanates are known to be detoxified by mammalian GSTs with great efficiency according to the reaction $R-N=C=S+GSH{\rightarrow}R-NH-C(SG)=S$ (Kolm et al. 1995).

Similar conjugating activities have also been demonstrated towards benzylisothiocyanate with GSTFs and GSTUs from wheat, maize and soybean (Cummins et al. 1997a; Skipsey et al. 1997; Dixon et al. 1998). In addition to forming part of the plant's chemical defences against allelochemicals produced by other plants, it is also tempting to speculate that GSTs are important in detoxifying fungal phytotoxins released during infection.

2.3.2.2 Binding and Transport of Bioactive Metabolites

GSTs have a long history of functioning as binding proteins, or ligandins, of biologically active ligands in mammalian liver. Ligands include tetrapyrrole bile salts as well as fatty acids and drugs (Hayes and Pulford 1995). The potential importance of a binding function for plant GSTs was reported in 1988, with the demonstration that uncharacterised GSTs from oats were strongly inhibited in their conjugating activity toward CDNB by tetrapyrroles at μM concentrations (Singh and Shaw 1988). Inhibition was particularly pronounced with porphyrin ligands. Subsequently, it was demonstrated that porphyrins gave non-competitive or mixed inhibition with purified maize GSTUs, suggesting that these ligands were not binding at the active site (Dixon et al. 1999). The significance of high-affinity binding of tetrapyrroles to GSTs has yet to be determined, but drawing parallels with what is known in animals, it is possible that such a ligandin function could serve to shuttle bio-

logically active porphyrins around the cell during the synthesis and degradation of organelles containing electron transport proteins (Hayes and Pulford
1995). In plants, the most abundant source of porphyrins is chlorophyll, with
disruptions in the biosynthesis of tetrapyrrole precursors giving rise to
photo-oxidative damage. However, the status of GSTs of ligandins of chlorophyll precursors is doubtful as there is little evidence that tetrapyrrole-bind-
ing GSTUs and GSTFs are expressed in the chloroplast where chlorophyll is
synthesised. However, it has been suggested that cytosolic GSTs bind porphyrins released during chloroplast degradation and that the resulting inhibition of these cytoprotective enzymes is a primary initiator of leaf senescence
(Singh and Shaw 1988).

In addition to binding porphyrins, several studies have demonstrated that
plant GSTs bind auxins (reviewed by Venis and Napier 1995) and cytokinins
(Gonneau et al. 1998). The evidence for hormone binding was obtained in the
first instance by identifying proteins that became photoaffinity labelled when
incubated with azido derivatives of auxins and cytokinins, respectively. With
both groups of hormones, binding is typically associated with GSTUs (Dixon
et al. 2002b). Whereas synthetic hormone derivatives bind tightly to the active
site of GSTs, the naturally occurring hormones give only weak and non-competitive inhibition (Droog et al. 1995). The physiological significance of the
weak binding of GSTs to auxins and cytokinins remains to be defined, though
it is perhaps significant that GSTUs are frequently induced by these hormones, or by the physiological conditions that favour their accumulation,
such as during cell division (Marrs 1996).

The most compelling evidence that GSTs play roles in ligand binding *in
planta* have come from defining the biochemical basis of the bronze II (*bz2*)
mutation in maize, which maps to a tau class GST gene (Marrs et al. 1995). In
these mutants oxidised anthocyanins aberrantly accumulate as bronze pigments in the cytoplasm as these flavonoid metabolites are not correctly transported into the vacuole. The mutation can be complemented by expression of
the BZ2 protein or distantly related phi class GSTs such as An9 from petunia
(Alfenito et al. 1998). Subsequent studies have shown that the An9 protein
binds flavonoids with high affinity at physiologically relevant concentrations,
while other GSTs that cannot complement bronze II do not (Mueller et al.
2000). In this case, it is proposed that BZ2 and An9 could serve multiple
important roles as ligandins in both preventing the uncontrolled release and
oxidation of unbound anthocyanins after their synthesis as well as promoting
their subsequent targeted delivery to vacuolar import protein transporters
(Mueller et al. 2000).

2.3.3 Alternative Functions for GSTs as Glutathione-Dependent Enzymes

2.3.3.1 Activity as Isomerases

It is becoming clear that many GSTs, especially those that have negligible or no activity as glutathione-conjugating transferases, can catalyse one of a number of important isomerase activities using glutathione as a coenzyme. It is even possible that GSTs originally evolved as isomerases and that transferase activities are "failed" isomerase activities, with glutathione lost to the highly reactive substrates in the course of catalysis. The most ubiquitous isomerase activity discovered so far for GSTs is the *cis–trans* isomerisation of maleylacetoacetate to fumarylacetoacetate, the penultimate reaction in the catabolism of tyrosine to fumarate and acetoacetate (Fig. 2.5). This reaction is catalysed by GSTZs, as has been shown for respective enzymes from humans, fungi (Fernández-Cañón and Peñalva 1998) and plants (Dixon et al. 2000). In the isomerisation, it is presumed that glutathione forms an addition product at the *cis* double bond, allowing the saturated bond to rotate to the *trans* configuration prior to removal of glutathione and desaturation. The crystal struc-

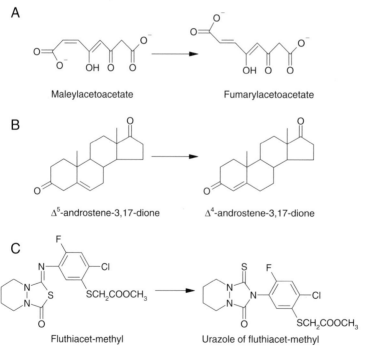

A

Maleylacetoacetate   Fumarylacetoacetate

B

Δ^5-androstene-3,17-dione Δ^4-androstene-3,17-dione

C

Fluthiacet-methyl   Urazole of fluthiacet-methyl

Fig. 2.5A–C. Isomerisation reactions catalysed by GSTs. **A** The *cis* to *trans* isomerisation of maleylacetoacetate to fumarylacetoacetate. **B** The Δ^5 to Δ^4 isomerisation of the double bond of androstene-3,17-dione. **C** The isomerisation of fluthiacet-methyl to the respective urazole

tures of zeta GSTs from *Arabidopsis* (Thom et al. 2001) and humans (Polekhina et al. 2001) have been determined; in both cases the enzymes fold and form dimers in a similar manner to other GSTs.

GSTs have also been shown to catalyse other isomerisation reactions of endogenous compounds. Mammalian sigma GSTs have prostaglandin D_2 synthase activity, catalysing the GSH-dependent isomerisation of prostaglandin H_2 to prostaglandin D_2 (Jowsey et al. 2001). Rat hepatic GSTs were able to catalyse the *cis–trans* isomerisation of 13-*cis*-retinoic acid to all-*trans*-retinoic acid (Chen and Juchau 1997), with human GSTP1–1 showing high activity with this substrate (Chen and Juchau 1998). GST-mediated isomerisation reactions are also important in the biosynthesis of steroid hormones such as testosterone and progesterone (Johansson and Mannervik 2001). Thus, human GSTA3–3 efficiently catalysed Δ^5 to Δ^4 double-bond isomerisations of steroids, such as of Δ^5-androstene-3,17-dione, with the enzyme being specifically expressed in steroid-producing tissues to catalyse this reaction (Fig. 2.5).

In plants the importance of GSTs in isomerisation reaction has been demonstrated with xenobiotic compounds. For example, the pro-herbicide fluthiacet-methyl was isomerised by velvetleaf GSTs to its urazole, which was active in inhibiting protoporphyrinogen oxidase (Shimizu et al. 1995; Fig. 2.5). In the absence of enzyme, this thiol-dependent reaction would only proceed at high pH. It is therefore likely that the GST was primarily catalysing the reaction by increasing the reactivity of the SH group of glutathione. Similarly, maize GSTs were found to isomerise thiadiazolidine-thiones to the active triazolidine-dithionesherbicides (Nicolaus et al. 1996; Uchida et al. 1997).

GSTs have also been shown to bioactivate *N*-(4-bromophenyl)-3,4,5,6-tetrahydroisophthalimide to *N*-(4-bromophenyl)-3,4,5,6-tetrahydrohydrophthalimide by an analogous isomerization reaction (Sato et al. 1997). Interestingly, in all these isomnerization reactions the GSTs are acting to bioactivate, rather than detoxify, their herbicide substrates.

2.3.3.2 Redox Functions

Some classes of GST do not have the requisite hydroxy group at the active site, in the form of a serine or tyrosine residue, to catalyse transferase reactions, but instead possess a sulphydryl group from a cysteine residue in the corresponding position. Such GSTs include the bacterial-specific beta class (Rossjohn et al. 1998), the mammalian omega class (Board et al. 2000) and the plant lambda class and dehydroascorbate reductases (DHARs) (Dixon et al. 2002a). Just as the DHARs have been shown to have a redox function, it is likely that the other classes mentioned have similar antioxidant activities. The DHARs act to recycle dehydroascorbate, formed from the spontaneous dis-

proportionation of monodehydroascorbate (MDA), to ascorbate, thus maintaining this important antioxidant in its reduced state. MDA can be reduced to
ascorbate by an NADPH-dependent enzyme, so DHARs may act as a backup
if the NADPH-dependent enzyme is unable to reduce MDA quickly enough.
In addition, ascorbate appears to be transported mainly as DHA, so DHARs
may be required to reduce DHA following transport (Noctor and Foyer 1998).

The lambda GSTs are a plant-specific class possessing a cysteine residue at
the active site, though their roles in redox function reactions have not been
determined. Lambda GSTs have been identified in maize and wheat based on
their strong induction of expression following treatment with herbicide
safeners (Hershey and Stoner 1991; Theodoulou et al. 1999). The co-induction
of these proteins with GSTUs and GSTFs with GPOX activity (see section
2.3.1.2) is consistent with an antioxidant function. However, unlike GSTUs
and GSTFs, members of the lambda class appear to be targeted to the chloroplast as well as the cytosol (Dixon et al. 2002a).

The distantly related omega GSTs are also able to reduce DHA, albeit with
lower efficiency than the plant DHARs. In addition to reducing DHA, the
human omega GST also catalyses the reduction of monomethylarsonate to
methylarsonous acid (Zakharyan et al. 2001). As such, the plant DHAR may
well play an analogous role in arsenic metabolism in plants. In addition, the
GSTOs, DHARs and GSTLs all catalyse thiol transfer reactions (Board et al.
2000; Dixon et al. 2002a). The exact significance of these thioltransferase reactions is unknown, though in other glutathione-dependent proteins such as the
glutaredoxins, this reaction removes glutathione from S-glutathionylated proteins. Since S-glutathionylation of proteins, also termed thiolation, is an
important method of regulating the activity of proteins with reactive sulphydryl groups (Klatt and Lamas 2000), both DHARs and GSTLs may have
important roles in regulating protein thiolation, a reaction which is known to
occur during oxidative stress.

2.3.3.3 Roles in Signalling

As described above, GSTs have long been known to be induced by various cellular events via diverse signalling pathways. Recent work has shown that in
some cases, GSTs may be involved in the signalling pathway itself, helping to
regulate other genes. Most of this work has been performed in non-plant systems, but they serve to illustrate the diverse and surprising roles of GSTs and
indicate possible functions for the plant enzymes. In addition very recent
studies have shown that a GSTU is essential in the UV-signalling in parsley,
suggesting that these proteins also play important roles in plant signalling
(Loyall et al. 2000).

The best studied GST involved in signalling is the human pi class GST
hGSTP. Under normal conditions, this enzyme has been shown to bind jun

kinase, inhibiting its phosphorylating activity. Stress invoked by treatment with UV or H_2O_2 causes multimerisation of hGSTP, which ameliorates its inhibition of jun kinase. This in turn activates a signalling cascade resulting in the expression of cytoprotective genes (Adler et al. 1999). Other examples include the stringent starvation protein SspA from *E. coli*, which is a member of the GST superfamily and which binds to RNA polymerase under conditions of starvation, leading to large scale changes in gene expression (Williams et al. 1994). Similarly, yeast Ure2, a prion protein with structural similarity to GSTs, is involved in the regulation of nitrogen metabolism (Bousset et al. 2001).

GSTs can also influence signalling in a less direct way, by influencing levels of its substrates and their products; a number of such compounds are known to modulate gene expression. For example, mouse GSTA4-4 efficiently metabolises lipid peroxides, which have roles as signalling molecules (Singh et al. 2002). Similarly, GSH is important in maintaining the redox state of cells, and any perturbation of redox status, due to excess ROS formation, for example, initiates signalling events to induce antioxidant protective systems (Noctor and Foyer 1998). Therefore, any GST that uses GSH at a rate sufficient to cause an appreciable decrease in GSH concentration is likely to give rise to a signalling response.

In plants, a GST in parsley has recently been shown to have a role in the response to UV irradiation in parsley (Loyall et al. 2000). Following UV treatment, transcripts encoding the enzyme chalcone synthase (CHS) are induced after about 3 h, which subsequently leads to the increased production of protective flavonoids. Searches for genes induced prior to CHS induction identified the tau class GST PcGST1, which was induced 35-fold after 2 h of UV treatment. Consistent with a role for this GST in signalling, constitutive over-expression of PcGST1 decreased the induction lag time of CHS, while addition of glutathione to cells overexpressing PcGST1 induced CHS in the absence of UV treatment. While the evidence for a signalling role for PcGST1 is good, the mechanism by which it induces CHS expression remains undetermined, though since PcGST1 required glutathione for signalling activity, the route of signalling may well involve the control of cellular redox status.

2.4 Roles for Plant GSTs in Xenobiotic Metabolism

GSTs are now implicated in herbicide selectivity in maize, sorghum, rice, soybean and wheat, with a variety of herbicide chemistries including chloro-*s*-triazines, chloroacetanilides, thiocarabamates, diphenyl ethers and selected aryloxyphenoxypropionates and sulphonyl ureas. As such, their role in herbicide metabolism and selectivity has been the subject of a recent detailed review (Edwards and Dixon 2000). Instead of re-reviewing the herbicide metabolising roles of these enzymes, in this section we will concentrate on those aspects of GST-mediated detoxification of xenobiotics that tell us something about the importance of these enzymes in the ecophysiology of plants.

2.4.1 Involvement of GSTs in Herbicide Metabolism and Selectivity

GSTs are frequently cited as classic mediators of the selectivity of herbicides between crops and weeds, with the associated rapid enzymic glutathione conjugation of the herbicide protecting against herbicide injury in the crop, while the metabolism of the herbicide in the competing weed is slower, resulting in greater phytotoxicity. In fact this perceived wisdom is largely derived from fragmented literature and has only recently been comprehensively addressed. In comparative studies where GST activities toward herbicides and associated rates of glutathione conjugation of herbicides were compared in crops and associated weeds in both maize (Hatton et al. 1996) and soybean (Andrews et al. 1997b), the levels of detoxifying GSTs were commonly ten-fold higher in the crop than in the weeds. However, this distinction was less obvious when generic model GST substrates such as CDNB were used. It was also interesting to note that contrary to earlier studies, it was the level of GST present, rather than the availability of glutathione (maize) or homoglutathione (soybean) that was the primary determinant of the rate of herbicide detoxification. In many respects, these studies pose more questions than they answer. Why do crops contain higher levels of herbicide detoxifying GSTs than non-domesticated weeds? In the few studies that have addressed this question, it appears that crops and weeds contain a similar diversity of herbicide-detoxifying GSTs. Thus, *Setaria faberi*, which is a problem weed in maize, contained an array of herbicide-detoxifying GSTs, including an isoenzyme that closely resembled the major GSTF in maize (Hatton et al. 1999). Instead, the major difference in the GST content of the two species was their relative level of expression, with the herbicide detoxifying isoenzymes being some 20-fold more abundant in maize than in *S. faberi*. Similar insights into the mechanism whereby GSTs define herbicide selectivity have been determined when the role of these enzymes in herbicide resistance in weeds have been examined. In *Abutilon theophrasti*, the resistance of a biotype toward the chloro-*s*-triazine atrazine was found to be due to the selective enhanced expression of isoenzymes that detoxified this herbicide (Anderson and Gronwald 1991). Similarly, in black-grass (*Alopecurus myosuroides*) populations showing cross resistance to phenylurea and aryloxyphenoxypropionate herbicides contained higher levels of GSTs than a herbicide-susceptible population (Cummins et al. 1997b; Reade and Cobb 1999). In particular, it was determined that a specific GSTF with high GPOX activity was expressed much more highly in the herbicide-resistant population with this isoenzyme protecting the weed from the oxidative damage invoked by herbicide injury (Cummins et al. 1999). From these studies we can conclude that it is the regulation of GST expression that is the primary determinant of herbicide tolerance in a plant, rather than the inherent genetic diversity in the genes encoding isoenzymes with differing detoxifying activities. In major crops such as maize, wheat, soybean, rice and sorghum the GSTs are all highly expressed, while in undomesticated weeds expression is normally much lower (Edwards and Dixon 2000). The reasons for the amplified expression of GSTs in many crops is a mystery, but is presumably linked to desirable agronomic traits selected for in the past in the course of domestication. In some cases we know the route whereby key herbicide detoxifying GSTs have originated in crops. In the case of hexaploid wheat (genome AABBDD), the GST with key roles in determining the selectivity of the herbicides dimethenamid and fenoxaprop has been exclusively donated from the DD progenitor *Triticum tauschii* (Riechers et al. 1997; Thom et al. 2002). However, perhaps more importantly we do not know the genetic origins of the regulatory switch(es) that cause the constitutively high levels of expression of GSTs in crops.

2.4.2 Herbicide Safeners and GST Regulation

Herbicide safeners are compounds that enhance herbicide tolerance in cereal crops largely by increasing the expression of detoxifying enzymes including GSTs. As well as enhancing GSTs, safeners also increase the synthesis of glutathione and the levels of ATP-binding cassette (ABC) transporter proteins that deposit the herbicide conjugates in the vacuole, thereby co-ordinately increasing the capacity of the glutathione conjugation detoxification pathway (Davies and Caseley 1999). Increased GST activity results from both increasing the expression of selected constitutive enzymes as well as the appearance of novel isoenzymes. In all cases reported to date, GST enhancement is due to increased gene transcription. Based on what is known of the regulation of *gst* genes it is likely that induction results from the binding of transcription factors to the *as-1* activator elements of the respective gene promoter (see section 2.2.2). However, the upstream events by which safeners are recognised to initiate signalling are unknown. In one model safeners act by initiating oxidative stress, which then leads to GST induction as an antioxidant response (Davies and Caseley 1999). However, if this is the case then the oxidative stress must be very restricted as modern safeners do not cause overt chemical damage or the accumulation of ROS (Cummins et al. 1999). In another model the parent safeners are recognised by a receptor protein. In support of this, structure activity studies with dichloroacetamide safeners in maize have demonstrated that these compounds interact with specific receptors (Walton and Casida 1995), and subsequently a safener binding protein (Saf BP) was isolated and cloned (Scott-Craig et al. 1998). The SafBP protein resembled an *O*-methyltransferase in sequence, though no functional catalytic or signalling activity was reported.

It has also been suggested that safeners require bioactivation to yield the active inducing agent (Davies and Caseley 1999). Interestingly, the maize safener benoxacor is rapidly metabolised by GSTs to a series of glutathione conjugates and derived metabolites that could function as the active signalling agent(s) (Miller et al. 1996). Whatever the mechanism of safener recognition, it is clear that they are activating a signalling pathway that originated to respond to biotic stress. One clue to the function of the safener response is that many of the proteins induced have antioxidant functions. For example, the GSTs that are safener responsive in maize and wheat have activities as GPOXs (Cummins et al. 1999; Sommer and Böger 1999). Such an hypothesis would also account for the safener-mediated accumulation of the antioxidant glutathione (Davies and Caseley 1999). If this is the case then a review of the literature would suggest that the interaction of the safener in the signalling cascade occurs downstream from oxidative stress and that there are several parallel pathways that also interact with signalling compounds such as plant hormones (see Sect. 2.2.2).

Safening of GSTs is conventionally only associated with large-grained cereal crops. In wheat the inducibility of GSTs has been found to be dependent on the genome type, with tetraploid AABB progenitors being less responsive than *Triticum* species with the AA, DD or hexaploid genomes (Riechers et al. 1996). Although the GST induction by safeners in wheat, maize, sorghum and rice is associated with increased herbicide tolerance, GSTs are also enhanced in competing grass weeds such as black-grass (Cummins et al. 1999). For reasons unknown, the safener induction of GSTs in weeds does not give similar levels of increased protection against herbicide injury in weeds as seen in cereal crops. Similarly, safeners induce GSTs in soybean and peas without increasing herbicide tolerance (Edwards and Dixon 2000). Safener-inducible signalling pathways seem to be conserved in monocots and dicots. Studies in which the promoter of a safener-inducible maize GSTL driving the expression of a β-glucuronidase (GUS) was introduced into *Arabidopsis* demonstrated that GUS expression was strongly enhanced following safener application (De Veylder et al. 1997). In the absence of safener the reporter gene was not expressed, with differing safeners giving distinct tissue-specific patterns of expression.

2.5 Future Goals in Understanding the Functions of Plant GSTs

From this short review it is clear that although there have been 3 decades of work on the importance of GSTs in herbicide metabolism in plants and numerous reports of their genes being upregulated in response to diverse stress treatments, we still have much to learn about these proteins and their functions in the metabolism of both endogenous and xenobiotic compounds.

To conclude, we now consider some of the latest developments in the field and future applications for GSTs.

2.5.1 GSTs in Transgenic Plants

Considering the long-standing links between GSTs and the tolerance of plants to biotic and abiotic stress, there are remarkably few reports of the respective coding sequences being over-expressed or used in antisense studies in transgenic plants. In contrast there are many reports of the promoters of inducible GSTs being used to drive the expression of reporter genes to study the regulation of these genes (see Sect. 2.2.2). Workers at Syngenta have reported the effects of the over-expression of a safener-inducible GSTF from maize in tobacco, with the transgenic plants showing increased tolerance to both chloroacetanilide and thiocarbamate herbicides (Jepson et al. 1997). Based on the increased tolerance of transgenic tobacco over-expressing an endogenous GSTF to salt and chilling stress (Roxas et al. 1997), it would also be interesting to determine whether or not the *Zm*GSTF transformed plants showed enhanced tolerance to biotic stresses. Such a careful cross-examination of phenotype in these transgenic plants relating traits in herbicide detoxification and stress tolerance and endogenous metabolism will be very important in determining why high GST expression has been selected for in the course of crop domestication.

The over-expression of GSTs in transgenic plants also suggests a method for engineering useful xenobiotic metabolising traits into species used in phytoremediation. In the case of plant GSTs this could be achieved by transferring genes encoding particularly active enzymes or increasing the expression of endogenous GSTs through the ectopic expression of regulatory transcription factors. Such modified GST expression could both increase the rates of GST-mediated detoxification as well as providing the plant with an increased ability to withstand oxidative stress resulting from the absorption of pollutants. It would also be of interest to express bacterial GSTs in plants, as these enzymes have dehalogenating activities not found in the corresponding plant enzymes that would be very useful in phytoremediating organochlorines (Vuilleumier and Pagni 2002). Having modified the capacity of the transgenic plant to detoxify pollutants using GSTs, it may then also be desirable to increase the expression of the ABC transporters responsible for the rapid removal of the inhibitory conjugates from the cytosol and into the vacuole (Rea et al. 1998)

2.5.2 GSTs as Targets for Functional Genomics

The completion of the *Arabidopsis* genome project together with large scale EST projects as described for the GSTs of soybean and maize (McGonigle et al.

2000) have given us an unprecedented knowledge of the diversity of the GST gene family. The new challenge is to identify how all the respective gene products function, and in the case of the GSTs, this is a major undertaking. The available evidence tells us that through changes in active site chemistry and structure, GSTs have evolved multiple functions using glutathione in catalysis as both coenzyme and co-substrate as well as having non-enzymic roles in binding and transport of biomolecules and in signalling (Table 2.1). A summary of how these GSTs with different roles may interact in the plant cell is

Table 2.1. GST classes, distribution, active site residue and function. With the exception of the lambda class GSTs, which are expressed as monomers, all other GST classes are functionally active as the respective dimeric proteins, composed of identical subunits, or subunits derived from the same class of GST

GST class	Distribution[a]	Active site residue	Functions
Alpha	V (Armstrong 1997)	Tyrosine	Steroid synthesis, detoxification
Mu	V (Armstrong 1997)	Tyrosine	Detoxification
Pi	V (Armstrong 1997), I (Engle et al. 2001)	Tyrosine	Detoxification, signalling
Sigma	V (Armstrong 1997), I (Chelvanayagam et al. 2001)	Tyrosine	Prostaglandin synthesis
Theta	V (Armstrong 1997), I (Chelvanayagam et al. 2001), P (Edwards et al. 2000)	Serine	Detoxification of oxidatively modified metabolites
Beta	B (Rossjohn et al. 1998)	Cysteine	?
Phi	P (Edwards et al. 2000)	Serine	Stress response?
Tau	P (Edwards et al. 2000)	Serine	Stress response?
Zeta	V (Fernández-Cañón and Peñalva 1998), I (Chelvanayagam et al. 2001), F (Fernández-Cañón and Peñalva 1998), P (Edwards et al. 2000)	Serine	Tyrosine catabolism, dechlorination
Omega	V (Board et al. 2000), I (Chelvanayagam et al. 2001)	Cysteine	Reductase/disulphide exchange?
Lambda	P (Dixon et al. 2002a)	Cysteine	Reductase/disulphide exchange?
Delta	I (Chelvanayagam et al. 2001)	Serine	Detoxification?
Kappa	V (Armstrong 1997)	?	? (mitochondrial)

[a] *V* Vertebrates, *I* invertebrates, *P* plants, *F* fungi, *B* bacteria.

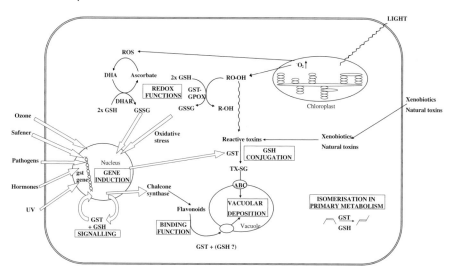

Fig. 2.6. A summary of the multifunctional roles ascribed to plant GSTs to date. *ABC* ATP-binding cassette transporter; *CHS* chalcone synthase; *DHA* dehydroascorbate; *DHAR* dehydroascorbate reductase; *GSH* glutathione (reduced); *GST* glutathione transferase; *GSSG* glutathione (oxidised)

shown in Fig. 2.6. The difficulties in characterising such multifunctional proteins is further compounded by redundancy, whereby several GSTs, even from different classes, can carry out similar functions. Taking *Arabidopsis* as an example, it will therefore be necessary to carry out multiple *gst* gene knockouts allied with detailed biochemical and physiological phenotyping to identify gene function. Such analyses may require a detailed ecological perspective, in that while some GSTs may have clearly identifiable essential functions that give a defined phenotype following knockout, in other cases the expression of multiple GSTs in plants may only give marginal improvements in growth, stress tolerance and seed set measurable only over several generations. In view of the independent evolution of the major classes of GSTs, it will also be intriguing to determine the similarities and differences in how plants and animals use these proteins in stress tolerance.

Acknowledgements. The authors acknowledge the financial support of the Biotechnology and Biological Sciences Research Council, which has contributed to the work described in this chapter.

References

Anonymous (2001) The *Arabidopsis* information resource. http://www.arabidopsis.org

Adler V, Yin ZM, Tew KD, Ronai Z (1999) Role of redox potential and reactive oxygen species in stress signaling. Oncogene 18:6104–6111

Alary J, Bravais F, Cravedi JP, Debrauwer L, Rao D, Bories G (1995) Mercapturic acid conjugates as urinary end metabolites of the lipid peroxidation product 4-hydroxy-2-nonenal in the rat. Chem Res Toxicol 8:34–39

Alfenito MR, Souer E, Goodman CD, Buell R, Mol J, Koes R, Walbot V (1998) Functional complementation of anthocyanin sequestration in the vacuole by widely divergent glutathione *S*-transferases. Plant Cell 10:1135–1149

Alla MMN (1995) Glutathione regulation of glutathione *S*-transferase and peroxidase activity in herbicide-treated *Zea mays*. Plant Physiol Biochem 33:185–192

Anderson MP, Gronwald JW (1991) Atrazine resistance in a velvetleaf (*Abutilon threophrasti*) biotype due to enhanced glutathione *S*-transferase activity. Plant Physiol 96:104–109

Andrews CJ, Jepson I, Skipsey M, Townson JK, Edwards R (1997a) Sequence of a glutathione transferase (accession No. Y10820) from soybean with activity towards herbicides. Plant Physiol 113:1005

Andrews CJ, Skipsey M, Townson JK, Morris C, Jepson I, Edwards R (1997b) Glutathione transferase activities toward herbicides used selectively in soybean. Pestic Sci 51:213–222

Armstrong RN (1997) Structure, catalytic mechanism, and evolution of the glutathione transferases. Chem Res Toxicol 10:2–18

Bartling D, Radzio R, Steiner U, Weiler EW (1993) A glutathione *S*-transferase with glutathione-peroxidase activity from *Arabidopsis thaliana*. Eur J Biochem 216:579–586

Barz W, Mackenbrock U (1994) Constitutive and elicitation induced metabolism of isoflavones and pterocarpans in chickpea (*Cicer arietinum*) cell suspension cultures. Plant Cell Tissue Organ Cult 38:199–211

Board PG, Coggan M, Chelvanayagam G, Easteal S, Jermiin LS, Schulte GK, Danley DE, Hoth LR, Griffor MC, Kamath AV, Rosner MH, Chrunyk BA, Perregaux DE, Gabel CA, Geoghegan KF, Pandit J (2000) Identification, characterization, and crystal structure of the omega class glutathione transferases. J Biol Chem 275:24798–24806

Boersma MG, Vervoort J, Szymusiak H, Lemanska K, Tyrakowska B, Cenas N, Segura-Aguilar J, Rietjens IMCM (2000) Regioselectivity and reversibility of the glutathione conjugation of quercetin quinone methide. Chem Res Toxicol 13:185–191

Booth J, Boyland E, Sims P (1961) An enzyme from rat liver catalyzing conjugations with glutathione. Biochem J 79:516–524

Bousset L, Belrhali H, Melki R, Morera S (2001) Crystal structures of the yeast prion Ure2p functional region in complex with glutathione and related compounds. Biochemistry 40:13564–13573

Chelvanayagam G, Parker MW, Board PG (2001) Fly fishing for GSTs: a unified nomenclature for mammalian and insect glutathione transferases. Chem Biol Interact 133:256–260

Chen H, Juchau MR (1997) Glutathione *S*-transferases act as isomerases in isomerization of 13-*cis*-retinoic acid to all-*trans*-retinoic acid in vitro. Biochem J 327:721–726

Chen H, Juchau MR (1998) Recombinant human glutathione S-transferases catalyse enzymic isomerization of 13-cis-retinoic acid to all-trans-retinoic acid in vitro. Biochem J 336:223–226

Cnubben NHP, Rietjens IMCM, Wortelboer H, van Zanden J, van Bladeren PJ (2001) The interplay of glutathione-related processes in antioxidant defense. Environ Toxicol Pharmacol 10:141–152

Cole DJ, Dixon DP, Cummins I, Edwards R (1998) Identification and characterisation of tau class glutathione transferases in maize and wheat responsible for herbicide detoxification. Abstr Am Chem Soc 216:095-AGRO

Cummins I, Cole DJ, Edwards R (1997a) Purification of multiple glutathione transferases involved in herbicide detoxification from wheat (*Triticum aestivum* L.) treated with the safener fenchlorazole-ethyl. Pestic Biochem Physiol 59:35–49

Cummins I, Moss S, Cole DJ, Edwards R (1997b) Glutathione transferases in herbicide-resistant and herbicide-susceptible black grass (*Alopecurus myosuroides*). Pestic Sci 51:244–250

Cummins I, Cole DJ, Edwards R (1999) A role for glutathione transferases functioning as glutathione peroxidases in resistance to multiple herbicides in black-grass. Plant J 18:285–292

Cummins I, O'Hagan D, Jablonkai I, Cole DJ, Hehn A, Werck-Reichardt D, Edwards R (2003) Cloning, characterization and regulatin of a family of phi class glutathione transferases from wheat. Plant Mol Biol (in press)

Danielson UH, Mannervik B (1985) Kinetic independence of the subunits of cytosolic glutathione transferase from the rat. Biochem J 231:263–267

Davies J, Caseley JC (1999) Herbicide safeners: a review. Pestic Sci 55:1043–1058

De Veylder L, Van Montagu M, Inzé D (1997) Herbicide safener-inducible gene expression in *Arabidopsis thaliana*. Plant Cell Physiol 38:568–577

Dean JV, Devarenne TP (1997) Peroxidase-mediated conjugation of glutathione to unsaturated phenylpropanoids. evidence against glutathione S-transferase involvement. Physiol Plant 99:271–278

Dean JV, Gronwald JW, Eberlein CV (1990) Induction of glutathione S-transferase isozymes in sorghum by herbicide antidotes. Plant Physiol 92:467–473

Dirr H, Reinemer P, Huber R (1994) X-ray crystal structures of cytosolic glutathione S-transferases – implications for protein architecture, substrate recognition and catalytic function. Eur J Biochem 220:645–661

Dixon DP, Cummins I, Cole DJ, Edwards R (1998) Glutathione-mediated detoxification systems in plants. Curr Opin Plant Biol 1:258–266

Dixon DP, Cole DJ, Edwards R (1999) Dimerisation of maize glutathione transferases in recombinant bacteria. Plant Mol Biol 40:997–1008

Dixon DP, Cole DJ, Edwards R (2000) Characterisation of a zeta class glutathione transferase from *Arabidopsis thaliana* with a putative role in tyrosine catabolism. Arch Biochem Biophys 384:407–412

Dixon DP, Cole DJ, Edwards R (2001) Cloning and characterisation of plant theta and zeta class GSTs: implications for plant GST classification. Chem Biol Interact 133:33–36

Dixon DP, Davis BG, Edwards R (2002a) Functional divergence in the glutathione transferase super-family in plants: identification of two classes with putative functions in redox homeostasis in *Arabidopsis thaliana*. J Biol Chem 277:30859–30869

Dixon DP, Lapthorn A, Edwards R (2002b) Plant glutathione transferases. Genome Biol 3:reviews3004.1-reviews3004.10

Droog F (1997) Plant glutathione S-transferases, a tale of theta and tau. J Plant Growth Regul 16:95–107

Droog FNJ, Hooykaas PJJ, Van der Zaal BJ (1995) 2,4-Dichlorophenoxyacetic acid and related chlorinated compounds inhibit two auxin-regulated type-III tobacco glutathione S-transferases. Plant Physiol 107:1139–1146

Edwards R (1996) Characterisation of glutathione transferases and glutathione peroxidases in pea (*Pisum sativum*). Physiol Plant 98:594–604

Edwards R, Dixon DP (2000) The role of glutathione transferases in herbicide metabolism. In: Cobb AH, Kirkwood RC (eds) Herbicides and their mechanisms of action. Sheffield Academic Press, Sheffield, pp 38–71

Edwards R, Dixon DP, Walbot V (2000) Plant glutathione S-transferases: enzymes with multiple functions in sickness and in health. Trends Plant Sci 5:193–198

Engle MR, Singh SP, Nanduri B, Ji X, Zimniak P (2001) Invertebrate glutathione transferases conjugating 4-hydroxynonenal:CeGST 5.4 from *Caenorhabditis elegans*. Chem Biol Interact 133:244–248

Eshdat Y, Faltin Z, Ben Hayyim G (1997) Plant glutathione peroxidases. Physiol Plant 100:234–240

Fernández-Cañón JM, Hejna J, Reifsteck C, Olson S, Grompe M (1999) Gene structure, chromosomal location, and expression pattern of maleylacetoacetate isomerase. Genomics 58:263–269

Fernández-Cañón JM, Peñalva MA (1998) Characterization of a fungal maleylacetoacetate isomerase gene and identification of its human homologue. J Biol Chem 273:329–337

Flury T, Adam D, Kreuz K (1995) A 2,4-D-inducible glutathione S-transferase from soybean (*Glycine max*). Purification, characterization and induction. Physiol Plant 94:312–318

Frear DS, Swanson HR (1970) Biosynthesis of S-(4-ethylamino-6-isopropylamino-2-s-triazino) glutathione: partial purification and properties of a glutathione S-transferase from corn. Phytochemistry 9:2123–2132

Fujita M, Adachi Y, Hanada Y (1995) Glutathione S-transferases predominantly accumulate in pumpkin culture cells exposed to excessive concentrations of 2,4-dichlorophenoxyacetic acid. Biosci Biotech Biochem 59:1721–1726

Gonneau J, Mornet R, Laloue M (1998) A *Nicotiana plumbaginifolia* protein labeled with an azido cytokinin agonist is a glutathione S-transferase. Physiol Plant 103:114–124

Gronwald JW, Plaisance KL (1998) Isolation and characterization of glutathione S-transferase isozymes from sorghum. Plant Physiol 117:877–892

Habig WH, Pabst MJ, Jakoby WB (1974) Glutathione S-transferases. J Biol Chem 249:7130–7139

Hatton PJ, Dixon D, Cole DJ, Edwards R (1996) Glutathione transferase activities and herbicide selectivity in maize and associated weed species. Pestic Sci 46:267–275

Hatton PJ, Cummins I, Cole DJ, Edwards R (1999) Glutathione transferases involved in herbicide detoxification in the leaves of *Setaria faberi* (giant foxtail). Physiol Plant 105:9–16

Hayes JD, Mclellan LI (1999) Glutathione and glutathione-dependent enzymes represent a co-ordinately regulated defence against oxidative stress. Free Radical Res 31:273–300

Hayes JD, McMahon M (2001) Molecular basis for the contribution of the antioxidant responsive element to cancer chemoprevention. Cancer Lett 174:103–113

Hayes JD, Pulford DJ (1995) The glutathione S-transferase supergene family – regulation of GST and the contribution of the isoenzymes to cancer chemoprotection and drug-resistance. Crit Rev Biochem Mol Biol 30:445–600

Henderson CJ, Wolf CR, Kitteringham N, Powell H, Otto D, Park BK (2000) Increased resistance to acetaminophen hepatotoxicity in mice lacking glutathione S-transferase Pi. Proc Natl Acad Sci USA 97:12741–12745

Hershey HP, Stoner TD (1991) Isolation and characterization of cDNA clones for RNA species induced by substituted benzenesulfonamides in corn. Plant Mol Biol 17:679–690

Hinson JA, Forkert P-G (1995) Phase II enzymes and bioactivation. Can J Physiol Pharmacol 73:1407–1413

Itzhaki H, Woodson WR (1993) Characterization of an ethylene-responsive glutathione S-transferase gene cluster in carnation. Plant Mol Biol 22:43–58

Jakobsson PJ, Morgenstern R, Mancini J, Ford-Hutchinson A, Persson B (1999) Common structural features of MAPEG – a widespread superfamily of membrane associated proteins with highly divergent functions in eicosanoid and glutathione metabolism. Protein Sci 8:689–692

Jepson I, Lay VJ, Holt DC, Bright SWJ, Greenland AJ (1994) Cloning and characterization of maize herbicide safener-induced cDNAs encoding subunits of glutathione S-transferase isoforms I, II and IV. Plant Mol Biol 26:1855–1866

Jepson I, Holt DC, Roussel V, Wright SY, Greenland AJ (1997) Transgenic plant analysis as a tool for the study of maize glutathione transferases. In: Hatzios KK, (ed) Regulation of enzymatic systems detoxifying xenobiotics in plants. KluwerAcademic Publishers, Dordrecht, pp 313–323

Johansson A-S, Mannervik B (2001) Human glutathione transferase A3–3, a highly efficient catalyst of double-bond isomerization in the biosynthetic pathway of steroid hormones. J Biol Chem 276:33061–33065

Jowsey IR, Thomson AM, Flanagan JU, Mudock PR, Moore GBT, Meyer DJ, Murphy GJ, Smith SA, Hayes JD (2001) Mammalian class Sigma glutathione S-transferases: catalytic properties and tissue-specific expression of human and rat GSH-dependent prostaglandin D_2 synthases. Biochem J 359:507–516

Kampranis SC, Damianova R, Atallah M, Toby G, Kondi G, Tsichlis PN, Makris AM (2000) A novel plant glutathione S-transferase/peroxidase suppresses Bax lethality in yeast. J Biol Chem 275:29207–29216

Klatt P, Lamas S (2000) Regulation of protein function by S-glutathiolation in response to oxidative and nitrosative stress. Eur J Biochem 267:4928–4944

Kolm RH, Danielson UH, Zhang Y, Talalay P, Mannervik B (1995) Isothiocyanates as substrates for human glutathione transferases: structure-activity studies. Biochem J 311:453–459

Kulkarni AP, Sajan MP (1997) A novel mechanism of glutathione conjugate formation by lipoxygenase: a study with ethacrynic acid. Toxicol Appl Pharmacol 143:179–188

Lamoureux GL, Rusness DG (1986) Tridiphane [2-(3,5-dichlorophenyl)-2-(2,2,2-trichloroethyl)oxirane] an atrazine synergist: enzymic conversion to a potent glutathione S-transferase inhibitor. Pestic Biochem Physiol 26:323–342

Lamoureux GL, Rusness DG (1993) Glutathione in the metabolism and detoxification of xenobiotics in plants. In: De Kok LJ et al. (eds) Sulfur nutrition and assimilation in higher plants. SPB Academic Publishing, The Hague, pp 221–237

Lopez MF, Patton WF, Sawlivich WB, Erdjumentbromage H, Barry P, Gmyrek K, Hines T (1994) A glutathione S-transferase (GST) isozyme from broccoli with significant sequence homology to the mammalian theta-class of GSTs. Biochim Biophys Acta 1205:29–38

Loyall L, Uchida K, Braun S, Furuya M, Frohnmeyer H (2000) Glutathione and a UV light-induced glutathione S-transferase are involved in signaling to chalcone synthase in cell cultures. Plant Cell 12:1939–1950

Marrs KA (1996) The functions and regulation of glutathione S-transferases in plants. Annu Rev Plant Physiol Plant Mol Biol 47:127–158

Marrs KA, Alfenito MR, Lloyd AM, Walbot V (1995) A glutathione S-transferase involved in vacuolar tansfer encoded by the maize gene Bronze-2. Nature 375:397–400

Mauch F, Dudler R (1993) Differential induction of distinct glutathione S-transferases of wheat by xenobiotics and by pathogen attack. Plant Physiol 102:1193–1201

McGonigle B, Keeler SJ, Lau S-MC, Koeppe MK, O'Keefe DP (2000) A genomics approach to the comprehensive analysis of the glutathione S-transferase gene family in soybean and maize. Plant Physiol 124:1105–1120

Miller KD, Irzyk GP, Fuerst EP, McFarland JE, Barringer M, Cruz S, Eberle WJ, Föry W (1996) Identification of metabolites of the herbicide safener benoxacor isolated from suspension-cultured Zea mays cells 3 and 24 h after treatment. J Agric Food Chem 44:3335–3341

Mueller LA, Goodman CD, Silady RA, Walbot V (2000) AN9, a petunia glutathione S-transferase required for anthocyanin sequestration, is a flavonoid-binding protein. Plant Physiol 123:1561–1570

Neuefeind T, Huber R, Dasenbrock H, Prade L, Bieseler B (1997a) Crystal structure of herbicide-detoxifying maize glutathione S-transferase-I in complex with lactoylglutathione: evidence for an induced-fit mechanism. J Mol Biol 274:446–453

Neuefeind T, Huber R, Reinemer P, Knäblein J, Prade L, Mann K, Bieseler B (1997b) Cloning, sequencing, crystallization and X-ray structure of glutathione S-transferase-III from Zea mays var. mutin: a leading enzyme in detoxification of maize herbicides. J Mol Biol 274:577–587

Nicolaus B, Sato Y, Wakabayashi K, Böger P (1996) Isomerization of peroxidizing thiadiazolidine herbicides is catalyzed by glutathione S-transferase. Z Naturforsch 51 c:342–354

Noctor G, Foyer CH (1998) Ascorbate and glutathione: keeping active oxygen under control. Annu Rev Plant Physiol Plant Mol Biol 49:249–279

Pettigrew NE, Colman RF (2001) Heterodimers of glutathione S-transferase can form between isoenzyme classes pi and mu. Arch Biochem Biophys 396:225–230

Pflugmacher S, Schröder P, Sandermann H (2000) Taxonomic distribution of plant glutathione S-transferases acting on xenobiotics. Phytochemistry 54:267–273

Polekhina G, Board PG, Blackburn AC, Parker MW (2001) Crystal structure of maleylacetoacetate isomerase/glutathione transferase zeta reveals the molecular basis for its remarkable catalytic promiscuity. Biochemistry 40:1567–1576

Rea PA, Li ZS, Lu YP, Drozdowicz YM, Martinoia E (1998) From vacuolar GS-X pumps to multispecific ABC transporters. Annu Rev Plant Physiol Plant Mol Biol 49:727–760

Reade JPH, Cobb AH (1999) Purification, characterization and comparison of glutathione S-transferases from black-grass (Alopecurus myosuroides Huds) biotypes. Pestic Sci 55:993–999

Reinemer P, Prade L, Hof P, Neuefeind T, Huber R, Zettl R, Palme K, Schell J, Koelln I, Bartunik HD, Bieseler B (1996) 3-dimensional structure of glutathione S-transferase from Arabidopsis thaliana at 2.2-angstrom resolution – structural characterization of herbicide-conjugating plant glutathione S-transferases and a novel active-site architecture. J Mol Biol 255:289–309

Riechers DE, Yang K, Irzyk GP, Jones SS, Fuerst EP (1996) Variability of glutathione S-transferase levels and dimethenamid tolerance in safener-treated wheat and wheat relatives. Pestic Biochem Physiol 56:88–101

Riechers DE, Irzyk GP, Jones SS, Fuerst EP (1997) Partial characterization of glutathione S-transferases from wheat (Triticum spp.) and purification of a safener-induced glutathione S-transferase from Triticum tauschii. Plant Physiol 114:1461–1470

Rossini L, Frova C, Mizzi L, Gorla MS (1998) Alachlor regulation of maize glutathione S-transferase genes. Pestic Biochem Physiol 60:205–211

Rossjohn J, Polekhina G, Feil SC, Allocati N, Masulli M, Di Ilio C, Parker MW (1998) A mixed disulfide bond in bacterial glutathione transferase: functional and evolutionary implications. Structure 6:721–734

Roxas VP, Smith RK, Allen ER, Allen RD (1997) Overexpression of glutathione S-transferase/glutathione peroxidase enhances the growth of transgenic tobacco seedlings during stress. Nat Biotech 15:988–991

Ruscoe JE, Rosario LA, Wang TL, Gate L, Arifoglu P, Wolf CR, Henderson CJ, Ronai Z, Tew KD (2001) Pharmacologic or genetic manipulation of glutathione S-transferase P1-1 (GST pi) influences cell proliferation pathways. J Pharmacol Exp Ther 298:339–345

Sato Y, Böger P, Wakabayashi K (1997) The enzymatic activation of peroxidizing cyclic isoimide: a new function of glutathione S-transferase and glutathione. J Pestic Sci 22:33–36

Schröder P, Berkau C (1993) Characterization of cytosolic glutathione S-transferase in spruce needles 1. GST isozymes of healthy trees. Bot Acta 106:301–306

Scott-Craig JS, Casida JE, Poduje L, Walton JD (1998) Herbicide safener-binding protein of maize. Plant Physiol 116:1083–1089

Sheehan D, Meade G, Foley VM, Dowd CA (2001) Structure, function and evolution of glutathione transferases: implications for classification of non-mammalian members of an ancient enzyme superfamily. Biochem J 360:1–16

Sherratt PJ, Manson MM, Thomson AM, Hissink EAM, Green T, Hayes JD (1998) Increased bioactivation of dihaloalkanes in rat liver due to induction of class theta glutathione S-transferase T1–1. Biochem J 335:619–630

Shimabukuro RH, Swanson HR (1969) Atrazine metabolism, selectivity, and mode of action. J Agric Food Chem 17:199–205

Shimizu T, Hashimoto N, Nakayama I, Nakao T, Mizutani H, Unai T (1995) A novel isourazole herbicide, fluthiacet-methyl, is a potent inhibitor of protoporphyrinogen oxidase after isomerization by glutathione S-transferase. Plant Cell Physiol 36:625–632

Singh BR, Shaw RW (1988) Selective inhibition of oat glutathione S-transferase activity by tetrapyrroles. FEBS Lett 234:379–392

Singh SP, Janecki AJ, Srivastava SK, Awasthi S, Awasthi YC, Xia SJJ, Zimniak P (2002) Membrane association of glutathione S-transferase mGSTA4–4, an enzyme that metabolizes lipid peroxidation products. J Biol Chem 277:4232–4239

Singhal SS, Tiwari NK, Ahmad H, Srivastava SK, Awasthi YC (1991) Purification and characterization of glutathione S-transferase from sugarcane leaves. Phytochemistry 30:1409–1414

Skipsey M, Andrews CJ, Townson JK, Jepson I, Edwards R (1997) Substrate and thiol specificity of a stress-inducible glutathione transferase from soybean. FEBS Lett 409:370–374

Sommer A, Böger P (1999) Characterization of recombinant corn glutathione S-transferases isoforms I, II, III, and IV. Pestic Biochem Physiol 63:127–138

Somssich IE, Hahlbrock K (1998) Pathogen defence in plants – a paradigm of biological complexity. Trends Plant Sci 3:86–90

Takahashi Y, Hasezawa S, Kusaba M, Nagata T (1995) Expression of the auxin-regulated parA gene in transgenic tobacco and nuclear localization of its gene product. Planta 196:111–117

Tang J, Siegfried BD, Hoagland KD (1998) Glutathione-S-transferase and in vitro metabolism of atrazine in freshwater algae. Pestic Biochem Physiol 59:155–161

Theodoulou FL, Clark IM, Pallett KE, Hallahan DL (1999) Nucleotide sequence of Cla 30 (Acession No. Y17386), a xenobiotic-inducible member of the GST superfamily from Triticum aestivum L. Plant Physiol 119:1567

Thom R, Dixon D, Edwards R, Cole D, Lapthorn A (2001) Structure determination of zeta class glutathione transferase from Arabidopsis thaliana. Chem Biol Interact 133:53–54

Thom R, Cummins I, Dixon DP, Edwards R, Cole DJ, Lapthorn AJ (2002) Structure of a tau class glutathione S-transferase from wheat active in herbicide detoxification. Biochemistry 41:7008–7020

Uchida A, Iida T, Sato Y, Boger P, Wakabayashi K (1997) Isomerization of 3,4-dialkyl-1,3,4-thiadiazolidines and 3,4-alkylene- 1,3,4-thiadiazolidines by glutathione S-transferase. Z Naturforsch 52 c:345–350

Ulmasov T, Ohmiya A, Hagen G, Guilfoyle T (1995) The soybean GH2/4 gene that encodes a glutathione S-transferase has a promoter that is activated by a wide range of chemical agents. Plant Physiol 108:919–927

Uotila M, Gullner G, Komives T (1995) Induction of glutathione S-transferase activity and glutathione level in plants exposed to glyphosate. Physiol Plant 93:689–694

Venis MA, Napier RM (1995) Auxin receptors and auxin-binding proteins. Crit Rev Plant Sci 14:27–47

Vuilleumier S, Pagni M (2002) The elusive roles of bacterial glutathione S-transferases: new lessons from genomes. Appl Microbiol Biotechnol 58:138–146

Wagner U, Edwards R, Dixon DP, Mauch F (2002) Probing the diversity of the *Arabidopsis* glutathione S-transferase gene family. Plant Mol Biol 49:515–532

Walczak HA, Dean JV (2000) Vacuolar transport of the glutathione conjugate of *trans*-cinnamic acid. Phytochemistry 53:441–446

Walton JD, Casida JE (1995) Specific binding of a dichloroacetamide herbicide safener in maize at a site that also binds thiocarbamate and chloroacetanilide herbicides. Plant Physiol 109:213–219

Wendel A (1981) Glutathione peroxidase. Methods Enzymol 77:325–333

Wilce MCJ, Parker MW (1994) Structure and function of glutathione S-transferases. Biochim Biophys Acta 1205:1–18

Williams MD, Ouyang TX, Flickinger MC (1994) Glutathione S-transferase SspA fusion binds to *Escherichia coli* RNA polymerase and complements Δ-SspA mutation allowing phage P1 replication. Biochem Biophys Res Commun 201:123–127

Wolf AE, Dietz K-J, Schröder P (1996) Degradation of glutathione S-conjugates by a carboxypeptidase in the plant vacuole. FEBS Lett 384:31–34

Xiang CB, Miao Z-H, Lam E (1996) Coordinated activation of *as-1*-type elements and a tobacco glutathione S-transferase gene by auxins, salicylic acid, methyl-jasmonate and hydrogen peroxide. Plant Mol Biol 32:415–426

Yin Z, Ivanov VN, Habelhah H, Tew K, Ronai Z (2000) Glutathione S-transferase p elicits protection against H_2O_2-induced cell death via coordinated regulation of stress kinases. Cancer Res 60:4053–4057

Zakharyan RA, Sampayo-Reyes A, Healy SM, Tsaprailis G, Board PG, Liebler DC, Aposhian HV (2001) Human monomethylarsonic acid (MMA^V) reductase is a member of the glutathione-S -transferase superfamily. Chem Res Toxicol 14:1051–1057

3 Activated Oxygen Species in Multiple Stress Situations and Protective Systems

Ron Mittler and Barbara A. Zilinskas

3.1 The Chemistry of Activated Oxygen Species

The electron configuration of molecular oxygen is unusual. It has two unpaired electrons, each in a π^* orbital and having parallel spins, and thus it is a triplet molecule in the ground state. This is in contrast to most other molecules in the cell, which exist in the singlet ground state where all electrons have paired spins. Reactions between molecular oxygen and most molecules are therefore forbidden because of spin restriction. Molecular oxygen, however, can be converted to activated oxygen species by overcoming the spin restriction by a spin flip producing singlet oxygen (O_2^1), or by the addition of either one, two or three electrons to form, respectively, the superoxide radical (O_2^-), hydrogen peroxide (H_2O_2) or the hydroxyl radical (OH). Unlike molecular oxygen, these activated oxygen species (AOS) can be very reactive and are often referred to as reactive oxygen species (ROS). Cells must have effective mechanisms to remove excess AOS, particularly the most highly reactive hydroxyl radicals, to prevent oxidative damage to cellular components.

3.2 Physiological Processes Contributing to AOS Production

3.2.1 Photosynthesis and Photorespiration

Environmental conditions that limit CO_2 fixation can result in absorption of more light energy than can be used productively in photosynthesis. Plants have several means to dissipate excitation energy, and these have been the subject of several excellent reviews (Asada 1999, 2000; Niyogi 1999, 2000; Foyer and Noctor 2000). Among these are included the direct reduction of O_2 by electrons on the acceptor site of PS I, the so-called Mehler-peroxidase reaction or the water–water cycle. Basically, when the ratio of NADPH/NADP$^+$ is

Ecological Studies, Vol. 170
H. Sandermann (Ed.)
Molecular Ecotoxicology of Plants
© Springer-Verlag Berlin Heidelberg 2004

high, single electrons provided by Fe-S centers associated with the PS I reaction center are diverted to O_2, producing O_2^-. The superoxide can be metabolized by a thylakoid-associated isoform of superoxide dismutase (SOD) to produce H_2O_2, which in turn is reduced to water by a thylakoid-bound ascorbate peroxidase (APX) with the concomitant oxidation of ascorbic acid (AsA) to monodehydroascorbate (MDA). Ascorbic acid is regenerated from MDA by reduced ferredoxin in PS I. There is no net O_2 exchange, and the name water-water cycle derives from the fact that the electrons released from H_2O by PS II are used to reduce O_2, and then they are reoxidized to water by APX. Since ∆pH and ATP are generated, while NADPH is not, this pathway has also been called pseudocyclic electron transport or pseudocyclic photophosphorylation. The water-water cycle is important physiologically as it removes O_2^- and H_2O_2 before they can diffuse from the thylakoid-associated scavenging system. Should some O_2^- or H_2O_2 escape the thylakoid system, a stromal system of antioxidant enzymes can take over to remove the AOS before they cause damage (Asada 1999, 2000).

Photorespiration, as is the case with the Mehler-peroxidase reaction, provides an alternate sink for photosynthetic energy and overreduction of the electron transport chain that would otherwise lead to photoinhibition. Photorespiration is initiated when Rubisco oxygenates RuBP producing phosphoglycolate. NADPH and ATP are consumed, and three organelles (chloroplasts, mitochondria and peroxisomes) are involved in reactions that recycle some of the fixed carbon and nitrogen that would otherwise be lost. Many environmental factors affect the magnitude of the Rubisco oxygenase and associated photorespiratory reactions. These include drought stress, which closes stomata and limits CO_2, which then increases oxygenase activity and photorespiration, or high temperature where oxygenase activity is increased because of the kinetic properties of the enzyme (Foyer and Noctor 2000; Wingler et al. 2000). In the process of photorespiration, H_2O_2 is produced at high rates through the action of glycolate oxidase, which is located in the peroxisomes. It must be efficiently dispensed with by catalase; otherwise the consequences can be serious, as has been observed with catalase antisense plants (Willekens et al. 1997).

3.2.2 Respiration and Other Oxidases

Another source of AOS production is the mitochondrion. Two sites of superoxide have been identified. The first site is the flavoprotein of NADH dehydrogenase (Complex I), where superoxide is formed in a univalent reduction of O_2. The second site is at ubiquinone. When normal electron transport from ubiquinone to the cytochrome b/c_1 complex (Complex II) is disrupted, ubisemiquinone reacts with O_2 to form superoxide. As the ubiquinone pool becomes more reduced, the rate of this reaction increases, and thus, as with photosynthetic electron transport, overreduction of the mitochondrial electron transport components leads to superoxide production. It has been observed that high levels of ATP and respiratory substrates favor the formation of O_2^- (Boveris and Cadenas 1982; Purvis and Shewfelt 1993). Superoxide, so generated in the mitochondria, is converted to H_2O_2 by mitochondrial Mn SOD.

It has been suggested that the mitochondrial alternative oxidase may serve to reduce the potential of AOS production in mitochondria by serving as an alternate sink for electrons (Purvis and Shewfelt 1993). Alternative oxidase accepts electrons directly from the ubiquinone pool, thus preventing accumulation of reduced ubiquinone, which can serve as a source of electrons for superoxide production. In support of this hypothesis, transgenic tobacco cells that overexpress the alternative oxidase produce less AOS than do wild-type cells, and conversely, antisense cells suppressed in alternative oxidase have higher levels of AOS (Maxwell et al. 1999).

A number of other oxidases generate AOS in plant cells. These include peroxidases (Bolwell and Wojitaszek 1997), amine oxidases (Allan and Fluhr 1997) and a NADPH-dependent oxidase (Groom et al. 1996; Keller et al. 1998) that are located respectively in the cell wall, apoplast and plasma membrane and are thought to contribute to the "oxidative burst" observed as one of the first responses of plants when they encounter bacterial and fungal pathogens. Superoxide and/or hydrogen peroxide are produced in two bursts. The first occurs upon interaction of plants with both compatible and incompatible pathogens. The second, which is stronger and more prolonged, is observed only with incompatible interactions, and it usually leads to localized cell death through the hypersensitive response (Grant and Loake 2000).

3.3 Detoxification of AOS in Plants

3.3.1 Enzymatic Mechanisms

As the production of AOS is an inevitable by-product of normal aerobic metabolism, systems have evolved to remove them to avoid injury to plant cell components. These enzymes, together with a variety of non-enzymatic, low molecular weight antioxidants and several ancillary enzymes, effectively rid the cell of toxic oxygen species under normal growth conditions. Oxidative stress becomes apparent when the factors that contribute to the production of reactive oxygen species outweigh the ability of the antioxidant system to eliminate them. When this balance is tipped, biological damage ensues.

While both O_2^- and H_2O_2 are themselves dangerous, the very reactive and highly toxic hydroxyl radical, for which no enzymatic scavenger exists, can be generated in the presence of O_2^- and H_2O_2 as a result of the metal-catalyzed Haber-Weiss reaction:

$$O_2^- + Fe(III) \quad \rightarrow \quad O_2 + Fe(II)$$
$$\underline{Fe(II) + H_2O_2 \quad \rightarrow \quad Fe(III) + OH^- + \cdot OH}$$
$$O_2^- + H_2O_2 \quad \xrightarrow{Fe} O_2 + OH^- + \cdot OH (Net).$$

Therefore, it is critical that the concentration of O_2^- and H_2O_2 be kept low to avoid oxidative damage. This is accomplished in plant cells by the action of SOD, APX and catalase (CAT). Ferritin and other metal-binding proteins are also critical in preventing damage to cells from hydroxyl radicals.

3.3.1.1 Superoxide Dismutase

There are several types of SODs, but all share a common denominator; they are metalloenzymes found in all aerobic organisms, which catalyze the dismutation of superoxide radicals to O_2 and H_2O_2:

$$O_2^- + O_2^- + 2H^+ \rightarrow H_2O_2 + O_2.$$

There are three major families of SODs: CuZn SOD, Fe SOD and Mn SOD. In plants, the CuZn SODs are the most abundant and are located in the cytosol, chloroplasts and apoplast. Mn SOD is found in mitochondria, and Fe SOD, which exists in a limited number of plant species, is located in chloroplasts. The enzyme reacts with O_2^- at a rate limited only by diffusion, and it very effectively removes O_2^-. Under stress, SOD activity is induced, presumably to cope with the increase in AOS. Expression of SOD isozymes is differentially regulated in response to environmental conditions and developmental cues (Bowler et al. 1992; Scandalios 1997).

3.3.1.2 Ascorbate Peroxidase

Hydrogen peroxide, a product of the SOD-catalyzed reaction, is itself toxic to cells, and efficient scavenging systems must exist to rid the cell of H_2O_2. In plants, the major enzymes are CAT, which is confined to peroxisomes and glyoxysomes, and APX, which is found in or associated with chloroplasts, peroxisomes, glyoxysomes, mitochondria and the cytosol. Ascorbate peroxidase reduces H_2O_2 to H_2O with an electron provided by ascorbate:

$$2 \text{ Ascorbate} + H_2O_2 \rightarrow 2 \text{ Monodehydroascorbate} + 2H_2O + 2H^+.$$

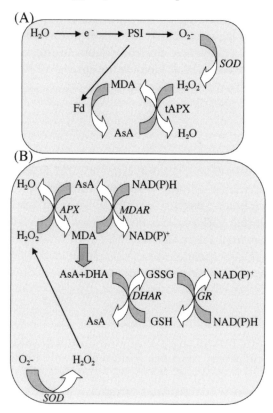

Fig. 3.1A, B. Pathways for scavenging O_2^- and H_2O_2 in chloroplasts. **A** Water–water cycle. **B** Ascorbate–glutathione cycle. Superoxide dismutase (*SOD*) acts as a first line of defense, converting O_2^- to H_2O_2. Ascorbate peroxidase (*APX*) then detoxifies H_2O_2, requiring ascorbate (*AsA*) as electron donor and a means to regenerate AsA. In the water–water cycle (A), electrons to regenerate AsA are provided by ferredoxin (*Fd*); in the ascorbate–glutathione cycle (B), NADPH provides the reducing power. *DHA* Dehydroascorbate; *DHAR* DHA reductase; *GR* glutathione reductase; *GSSG* oxidized glutathione; *MDA* monodehydroascorbate; *MDAR* MDA reductase

When this reaction is catalyzed by thylakoid-bound APX, the MDA so generated is reduced to ascorbate (AsA) with electrons provided by reduced ferredoxin (Fig. 3.1A). When other APX isozymes produce MDA, it is reduced back to AsA through the ascorbate-glutathione cycle (Fig. 3.1B). In this cycle, MDA can be directly reduced to AsA by monodehydroascorbate reductase (MDAR), using NAD(P)H as an electron donor:

$$MDA + NAD(P)H + H^+ \rightarrow Ascorbate + NAD(P)^+.$$

Alternatively, MDA can spontaneously disproportionate to AsA and dehydroascorbate (DHA):

$$2\,MDA + 2H^+ \rightarrow Ascorbate + DHA.$$

In this case, the DHA is reduced by dehydroascorbate reductase (DHAR), using glutathione (GSH) as electron donor; in turn, oxidized glutathione (GSSG) is reduced by glutathione reductase (GR), using NAD(P)H as electron donor (Foyer and Halliwell 1976; Nakano and Asada 1981).

Ascorbate peroxidase is a heme-containing protein that has a very high affinity for H_2O_2. The enzyme has a high specificity for AsA as electron donor, although the cytosolic isoform can use alternate electron donors. In contrast to the cytosolic isoform, the chloroplastic APX is labile in the absence of AsA (Asada 1999).

3.3.1.3 Catalase

In contrast to APX, which is found in many cell compartments, CAT is confined to peroxisomes and glyoxysomes where it functions respectively to remove H_2O_2 produced during photorespiration and β-oxidation of fatty acids. It also differs from APX in that CAT requires no reductant, has a very low affinity for H_2O_2 and has a very high Vmax. Catalase, which is a heme protein, must be present in very high concentration to effectively remove H_2O_2 (Willekens et al. 1995). It catalyzes the reaction:

$$2 H_2O_2 \rightarrow 2 H_2O_2 + O_2.$$

Studies on a barley CAT-deficient mutant (Kendall et al. 1983) and tobacco CAT-antisense plants (Chamnongpol et al. 1996; Willekens et al. 1997) demonstrated that CAT is required to deal with H_2O_2 generated during photorespiration. Under photorespiratory conditions, expression of APX and glutathione peroxidase (GPX) increased in the CAT-antisense plants; however, this was insufficient to compensate for the low levels of catalase (Willekens et al. 1997). A similar phenomenon holds true for APX-antisense plants, where induction of SOD, CAT and GR did not replace the need for APX (Rizhsky et al. 2002a).

3.3.1.4 Other Antioxidant Enzymes

In addition to the key enzymes noted above, other enzymes have recently been identified in plants that play an important role in reducing lipid peroxidation. These include phospholipid hydroperoxide glutathione peroxidase (GPX) and two-cysteine peroxiredoxin (2-CP). These enzymes reduce alkylhydroperoxides to their corresponding alcohols, thus limiting the likelihood of further lipid peroxidation. GPX is thought to be located in the cytosol and the chloroplast stroma, while 2-CP is associated with the stroma-side of the thylakoid membrane. Of lesser significance in H_2O_2 removal, these enzymes appear to be critical in reduction of organic peroxides (Eshdat et al. 1997; Baier and Dietz 1999).

3.3.2 Nonenzymatic Mechanisms

3.3.2.1 Water-Soluble Antioxidants

Two antioxidants that are present in millimolar concentrations in plant cells are ascorbic acid (AsA), known also as vitamin C, and glutathione (GSH). Vitamin C can directly quench singlet oxygen and superoxide and hydroxyl radicals, and thus it is critical in protecting plants against oxidative stress. It is also important in maintaining vitamin E in its reduced state by regenerating it from its α-tocopheryl radical. Ascorbate also serves as a substrate for APX and violaxanthin de-epoxidase; the latter enzyme is involved in the xanthophyll cycle, which is key in thermal dissipation of excess excitation energy in PS II. As might be predicted from AsA's role as an antioxidant, an AsA-deficient *Arabidopsis* mutant, identified in a screen for increased sensitivity to ozone, was shown to be more sensitive to a variety of environmental stresses (Smirnoff 2000).

Glutathione, in its reduced form, is a tripeptide, γ-glu-cys-gly. Glutathione contributes in several ways to antioxidant capability. First, it is a cofactor for DHAR in the ascorbate-glutathione pathway, providing electrons to reduce DHA to AsA. Alternately, at alkaline pH, AsA can be regenerated nonenzymatically from DHA with electrons donated directly from GSH. Second, like AsA, GSH can detoxify singlet oxygen and hydroxyl radicals. Lastly, it plays an important role in protecting thiol groups on sensitive stromal enzymes. Several studies have shown that GSH is synthesized in response to stress and that GSH content may be elevated constitutively in species that are subject to stressful growth conditions (Noctor and Foyer 1998).

3.3.2.2 Lipid-Soluble Antioxidants

Carotenoids and tocopherols are important antioxidants that are membrane associated. Tocopherols, of which α-tocopherol is the most abundant, chemically scavenge and physically quench singlet oxygen and superoxide and hydroxyl radicals, and in doing so, prevent lipid peroxidation. The tocopherols are also efficient terminators of lipid peroxidation chain reactions.

Carotenoids, which are bound to proteins, are particularly important in protecting chlorophyll through their ability to quench triplet chlorophyll and singlet oxygen. Carotenoids have also been demonstrated to significantly inhibit lipid peroxidation. Extensive studies on mutants deficient in specific carotenoids have provided evidence for the important role these antioxidants play in photoprotection (Niyogi 1999).

3.4 Production of AOS During Environmental Stress

Our atmosphere, containing about 20 % oxygen, poses a constant threat to all organisms. Many chemical and biochemical reactions that occur in cells during normal metabolism are accompanied by an unavoidable process of oxygen reduction and the formation of AOS. This continuous process of AOS production is countered by active removal of AOS via antioxidants that maintain the level of AOS within cells at a bearable level. Nevertheless, during environmental stress, the balance between AOS production and AOS detoxification may be shattered and the overall level of AOS can dramatically increase, threatening to injure and even kill cells. The rise in AOS production during environmental stress typically results from stress-related changes in cellular homeostasis that disrupt and uncouple normal metabolic pathways and divert electrons from these pathways to the reduction of oxygen and the formation of AOS (Allen 1995; Noctor and Foyer 1998; Dat et al. 2000; Mittler 2002).

Under optimal environmental conditions, cellular homeostasis is achieved by the coordinated action of many biochemical pathways. However, different pathways may have distinct molecular and biophysical properties, making them different in their dependence upon physical conditions. Thus, during events of sub-optimal conditions (stress), different pathways will be affected differently and their coupling, which makes cellular homeostasis possible, will be disrupted. A central example for this phenomenon is the coupling of the light and dark reactions of photosynthesis during different stresses, as described below.

3.4.1 Drought Stress

The absorption of light by the chlorophyll-containing antennae of the photosynthetic apparatus and the transfer of excitation energy to the photosynthetic centers and subsequent transfer of electrons through the thylakoid electron transport chain are not as sensitive to water loss as is the fixation of CO_2 by Rubisco during the dark reaction (Krause and Cornic 1987; Asada 1999). Although the light reaction can be affected by a considerable loss of water, e.g., desiccation, a loss of water on a much lower scale will result in stomatal closure. This will limit the supply of CO_2 for carbon fixation and inhibit the dark reactions of photosynthesis, i.e., the Benson-Calvin cycle and Rubisco. As a result, electrons supplied by the light reaction for CO_2 fixation will be used for AOS production instead of carbon fixation [i.e., the Mehler reaction and the water-water cycle (Asada 1999)]. In addition, in C3 plants during drought stress, the photorespiratory pathway is activated and H_2O_2 is produced in the peroxisomes. Thus, although some of the impact of drought stress is removed from the chloroplast, AOS are produced during this process in peroxisomes (Asada and Takahashi 1987; Dat et al. 2000).

3.4.2 Unfavorable Temperatures

The differential sensitivity of cellular pathways to adverse physical conditions may also result from the dependence of certain pathways on the function of membrane-embedded complexes and channels. Compared to soluble proteins and enzymes, these highly organized multisubunit complexes can be very sensitive to changes in temperature. Thus, heat shock, chilling or freezing injury may result in changes in membrane structure, the disassembly of these complexes and their inhibition. This may result in the formation of AOS via two different routes: (1) electrons that were supposed to flow between different subunits of the complex may be diverted into oxygen reduction, or (2) coupling of the membrane-dependent pathways with membrane-independent and temperature-tolerant pathways will be disrupted resulting in the channeling of energy into AOS formation. An example may come from studying the mitochondrial respiration pathway during heat shock. In this pathway NADH produced by the Krebs-cycle donates electrons to the mitochondrial electron transport chain. These are used to reduce oxygen to water and produce an electrochemical gradient. However, during heat shock it is thought that membrane-bound complexes are inhibited or disrupted, and electrons from NADH are used to reduce O_2 to AOS by different components of the uncoupled electron transport chain (Davidson and Schiestl 2001).

3.4.3 Salt Stress

Salt stress may affect plant cells via at least two different mechanisms: (1) it will result in the disruption of the osmotic balance of cells, somewhat similar to drought stress, and (2) it will result in the inhibition of different enzymes and pathways because of the toxicity of sodium ions. In either case the cellular homeostasis of cells will be disrupted and the rate of AOS production will increase. A direct example for the toxicity of AOS during salt stress comes from studying the response of catalase-deficient plants to this stress. These plants were found to be hypersensitive to salt stress (Willekens et al. 1997), suggesting that the photorespiratory pathway is activated in plants during salt stress in a similar manner to its activation during drought (Dat et al. 2000).

3.4.4 High Light

The effect of high light stress has been extensively studied in plants. High light is mostly a stress to shade-adapted plants or when it occurs together with other stresses such as drought or cold. During high light stress the photosynthetic apparatus is functioning at a very high rate, one that overcomes the rate of CO_2 fixation by Rubisco. AOS are thought to be produced directly by the photosynthetic apparatus at the PS I and PS II centers. These are scavenged by SOD and APX in the water-water cycle (Asada 1999). However, AOS produced at the PS II center can damage the D1 protein and lead to its degradation (photoinhibition; Powles 1984; Mullineaux and Karpinski 2002). Thus, high light stress is accompanied by a high turnover rate of this protein. The uncoupling of different components in the chloroplast electron transport chain that may result from D1 degradation may lead to over-excitation of antennae chlorophyll and the production of singlet oxygen. This AOS is scavenged by carotenoids found within the antennae. In *Arabidopsis* high light stress is accompanied by the induction of two isozymes of cytosolic APX (ApxI and ApxII; Mullineaux and Karpinski 2002). Interestingly, high light stress in this plant was not found to result in the induction of chloroplastic systems for AOS detoxification (Karpinski et al. 1997). It is therefore possible that AOS produced within the chloroplast diffuse into the cytosol and cause the induction of antiperoxidative enzymes in this compartment. However, it was not AOS but rather the redox status of the plastoquinone pool that was found to signal the induction of APXII during this stress (Mullineaux and Karpinski 2002). Moreover, it was found that high light stress applied to a single leaf of an *Arabidopsis* plant results in the generation of a systemic signal that travels to other parts of the plant and induces APXII expression in parts of the plant that were not subjected to high light stress (Karpinski et al. 1999).

3.4.5 Ozone

Ozone is an air pollutant that can penetrate plant tissues when the stomates are open. It induces the formation of various AOS upon interacting with the cell wall and plasma membrane of cells. Ozone stress was found to be accompanied by the induction of different AOS scavenging systems. In addition, it was found that the signal transduction pathway activated upon ozone stress overlaps with the signal transduction pathway activated upon pathogen recognition (Sandermann et al. 1998; Rao and Davis 1999). Because the plant-pathogen signal transduction pathways involve the production of AOS at the cell wall and plasma membrane of cells, it is possible that ozone and pathogens may overlap by producing similar AOS within similar compartments. Because of this overlap, at least part of the cell death induced by ozone may be the result of activating a programmed cell death pathway that is also activated by pathogens during the hypersensitive response (Rao and Davis 1999; Dat et al. 2000).

3.4.6 UV-B Radiation

UV-B radiation can lead to the generation of superoxide radicals by different photosensitizers found in plant cells. Accordingly, UV-B stress was found to result in the induction of antioxidative enzymes (Surplus et al. 1998). Interestingly, UV-B stress, similar to ozone stress, was found to be accompanied by the induction of pathogenesis related proteins, possibly as a result of H_2O_2 accumulation (Dat et al. 2000).

3.4.7 Heavy Metals

Heavy metals such as aluminum, cadmium, copper, zinc or iron were found to induce oxidative stress in plants. Because of their atomic properties, heavy metals can induce uncontrolled redox reactions that will lead to the oxidation of membranes and proteins and the disruption of cellular homeostasis. In addition, some heavy metals, e.g., iron, can directly interact with superoxide and peroxide ions to form hydroxyl radicals through the Haber-Weiss or Fenton reactions (Halliwell and Gutteridge 1989).

The level of AOS produced during the different stresses described above was directly measured in only a few examples (Dat et al. 2000); however, one relatively clear indication that AOS production is increased during these stresses can be found in the form of induction of AOS-scavenging mechanisms. Thus, almost all of the stresses described above are accompanied by an enhanced expression of antioxidative enzymes such as APX, CAT and SOD (Bowler et al. 1992; Allen 1995; Dat et al. 2000). These can serve as internal markers for AOS production. Moreover, the presence of some of these antioxidative enzymes was found to be critical for the stress tolerance of plants. Thus, mutants that lacked enzymes such as APX and CAT were found to be hypersensitive to different environmental stress conditions such as ozone, high light, salt stress, drought and pathogen attack (Orvar and Ellis 1997; Willekens et al. 1997; Mittler et al. 1999).

3.5 Production of AOS During Recovery from Stress

The removal of AOS produced during stress as a result of the disruption of cellular homeostasis appears to be critical for plant survival (Allen 1995; Noctor and Foyer 1998; Dat et al. 2000; Mittler 2002). However, AOS may also be produced during the recovery period that follows stress (Mittler and Zilinskas 1994).

In nature, plants are subjected to cycles of stress and relief from stress as the conditions within their ecosystem fluctuate. Recovery from a stressed, unbalanced metabolism to homeostasis is therefore a critical aspect of plant acclimation to changes in environmental conditions. The process of cellular recovery may require a coordinated change in the expression pattern of many genes belonging to different pathways. However, it is possible that not all cellular mechanisms will recover simultaneously. For example, cytosolic and chloroplastic pathways may recover at different rates resulting in the uncoupling of normally coupled metabolic processes. Such a situation may result in a transient, unbalanced metabolic state, albeit perhaps different from that which occurs during stress. Thus, the potential risk of producing toxic intermediate compounds such as AOS may also be present during recovery from stress. Successful recovery may, consequentially, depend upon the capability of defense systems to efficiently detoxify AOS that arise during this transient period.

It was previously found that during recovery from drought the expression of the antioxidative enzymes SOD and APX is induced (Mittler and Zilinskas 1994). This finding suggested that during recovery from drought plants may be subjected to an oxidative stress. The recovery of plants from other stresses such as cold or high light was also found to be accompanied by the induction of AOS-scavenging enzymes such as APX (Karpinski et al. 1997) and glutathione reductase (Stevens et al. 1997). An enhanced rate of AOS formation may therefore accompany the recovery of plants from different environmental stress conditions. Unfortunately, this aspect of plant biology received little attention in recent years, and it is not known what are the main producers of AOS during recovery from stress and what are the main pathways for AOS removal induced during this process.

3.6 Natural Conditions Within the Ecosystem, a Combination of Environmental Stresses

To date, very little is known about the molecular mechanisms involved in the response of plants to environmental conditions within their natural habitat. This knowledge may be critical because, unlike the controlled experimental conditions used in the laboratory to study the response of plants to environmental stress, a large number of different factors may affect the growth of plants within their natural ecosystem. For example, desert plants naturally growing within an arid dune ecosystem were found to be subjected to a combination of different stress conditions. These included high irradiance, low water availability, high or low temperature, high salt concentrations and nutrient deprivation (Mittler et al. 2001). Moreover, the conditions described above could change very rapidly within a few hours during the day, and more slowly but not less extensively between different seasons. They could vary from a temperature of 42°C, a relative humidity of 10% and a light intensity of 2,000 µmol m^{-2} s^{-1} during noon in a typical day of the dry season, to a sub-zero temperature during the night in the rainy season (Merquiol et al. 2002). These conditions are very different from the conditions used in the laboratory to study changes in gene expression in response to environmental stress.

As described above, each of the different components contributing to the stressful conditions that occur within the arid dune ecosystem, i.e., drought, heat stress, high light stress and chilling, may result in an enhanced rate of production of AOS. In addition, AOS can be formed upon wounding or herbivory with mediation by oxylipins and systemin (in Solanaceous plants) (Kessler and Baldwin 2002). However, little is known about how the combined effect of these different stresses on plant metabolism will alter the intracellular level of AOS. Studying the molecular mechanisms involved in the acclimation of plants to naturally occurring conditions may therefore unravel complex relations between known mechanisms and possibly reveal novel pathways and strategies that may enable plants to resist stressful conditions.

3.7 Effect of Multiple Stresses on Plant Metabolism: The Example of Desert Plants

Different environmental stress conditions may disrupt or alter different cellular pathways, leading to the activation of stress-specific responses that counter the effect of the particular stress. However, in the field, plants may be exposed to a combination of different stresses. Such a combination of stresses may require a new type of response that would not have been induced by only one of the different stress conditions. For example, the expression of dehydrin, elevated dramatically during drought, is only slightly elevated during a combination of drought and heat shock. In contrast, the expression of a number of different transcripts is specifically expressed during a combination of drought and heat shock. These include specific transcription factors (WRKY and ERTCA), small heat shock proteins and mitochondrial alternative oxidase (Rizhsky et al. 2002b). The combination of heat and drought stress may therefore have a very different effect on the metabolism of plants compared to each of the stresses alone. This type of response can be expected because drought stress and heat stress affect the plant differently and may inhibit different metabolic pathways. In contrast, stresses such as cold and drought or UV-B and pathogen infection were found to result in the induction of similar defense pathways, suggesting that some stresses may affect the metabolism of plants in a similar manner (Bowler and Fluhr 2000; Langebartels et al. 2000; Seki et al. 2001). Chapter 4 provides a detailed presentation of such similarities with respect to induced cell death.

To examine the mechanisms involved in the response of plants to multiple stresses, the C3 desert plant *Retama raetam* was used as a model (Fig. 3.2). A 2-year field study was conducted on the molecular responses involved in the acclimation of *R. raetam* plants, naturally growing within their ecosystem, to diurnal and seasonal changes in their environment. In order to survive their harsh environment, *R. raetam* plants use a combination of different avoidance and resistance strategies. They may use active defense mechanisms such as the enhanced expression of dehydrins, heat shock proteins and AOS-removal enzymes to resist stress. They may enter a state of overall metabolic suppression coupled with induction of defense genes, referred to as "dormancy," or

Fig. 3.2A–D. (in color at end of book) Dormant and non-dormant *Retama raetam* plants. **A** A non-dormant *R. raetam* plant (approximately 2 m high). **B** A dormant *R. raetam* plant (approximately 2.5 m high). **C** Stems of the non-dormant *R. raetam* plant shown in A. **D** Stems of the dormant *R. raetam* plant shown in B. The two plants were photographed at the same research site on the same day (in June 2000), about 20 min apart

they can enter an intermediate state of metabolic suppression resulting in the partial inhibition of photosynthesis and the induction of multiple defense enzymes (Mittler et al. 2001; Merquiol et al. 2002).

Dormancy in *R. raetam* is accompanied by a decrease in relative water content, the complete suppression of photosynthesis and photosynthetic genes and the suppression of many metabolic genes, apparently by a post-transcriptional mechanism. This state is reversible within 12–24 h following rainfall (Mittler et al. 2001). Intermediate metabolic suppression is mostly apparent during mid-day in non-dormant plants when the environmental conditions are the harshest. An example is shown in Figs. 3.3 and 3.4. As can be seen, the majority of photosynthetic activity is carried out during the morning hours (between 7:00   and 10:00 a.m.) and again during the afternoon hours (between 3:00 and 5:00 p.m.; Fig. 3.3). During the midday hours, i.e., between 11:00 a.m. and 3:00 p.m., when environmental conditions are the harshest, photosynthesis is suppressed and defense enzymes, such as heat shock pro-

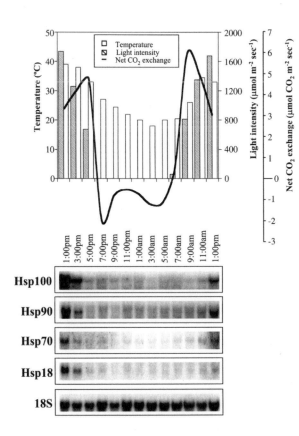

Fig. 3.3. Changes in the expression pattern of heat shock proteins (*HSPs*) and in CO_2 assimilation (a measure of photosynthetic activity) during daytime and night in the desert plant *R. raetam* growing naturally within an arid dune ecosystem (the Nizzana site). *Top* Graph showing the changes in temperature and light intensity and the changes in the rate of CO_2 assimilation as recorded every 2 h from 1:00 p.m. of one day to 1:00 p.m. of the next day (at the beginning of September 2000). *Bottom* RNA gel blot analysis of the expression pattern of four different cytosolic HSPs (*HSP100, 90, 70* and *18*) in samples obtained at the different times as described above. RNA gel blots were first hybridized with the different HSPs probes and then with a probe for 18S rRNA. From Merquiol et al. 2002

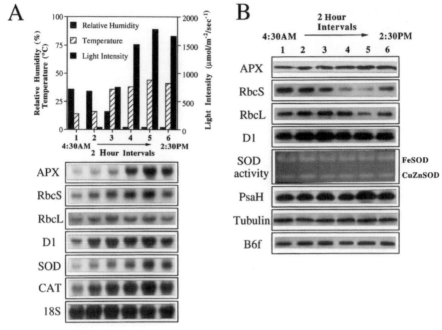

Fig. 3.4. Changes in the expression pattern of transcripts encoding AOS-scavenging enzymes and photosynthetic proteins and changes in proteins involved in photosynthesis in the desert plant *R. raetam* growing naturally within an arid dune ecosystem (the Nizzana site). **A** *Top* Changes in environmental parameters between 4:30 a.m. and 2:30 p.m. at the research site. *Bottom* Changes in the expression pattern of transcripts encoding AOS scavenging enzymes (*APX, SOD* and *CAT*) and proteins involved in photosynthesis (*RbcS, RbcL* and *D1*) in samples obtained from plants at the different time points indicated above. The *full bars* for relative humidity and light intensity appear on the left- and right-hand side, respectively, of the *hatched bar* for temperature. **B** Changes in the expression pattern of AOS-scavenging enzymes (*APX* and *SOD*) and proteins involved in photosynthesis (*RbcS, RbcL, D1 PsaH* and *B6f*) in samples obtained from plants at the different time points indicated in A (*top*). During the stressful midday hours, the expression of transcripts encoding AOS-scavenging enzymes is shown to be induced (A), and the level of Rubisco protein (*RbcL* and *RbcS*; B) is shown to be suppressed (B), suggesting a complex mode of transcriptional and post-transcriptional regulation of gene expression (from Mittler et al. 2001). *B6f* a subunit of the cytochrome B6f complex of the thylakoid membrane; *D1* a subunit of the photosynthetic reaction center complex II; *PsaH* a subunit of the photosynthetic reaction center complex I; *RbcS* the small subunit of Rubisco; *RbcL* the large subunit of Rubisco)

teins (Fig. 3.3; Merquiol et al. 2002) and AOS-removal enzymes (Fig. 3.4; Mittler et al. 2001), are induced. This induction is accompanied by a decrease in the level of some proteins involved in photosynthesis (Fig. 3.4B).

The photosynthetic apparatus is considered to be one of the main cellular producers of AOS during stress (Asada and Takahashi 1987). It is therefore reasonable to assume that the suppression of photosynthetic activity during metabolic suppression (dormancy or intermediate suppression) is likely to result in a considerable reduction in the rate of AOS formation. These avoidance mechanisms may be viewed therefore as a defense strategy aimed at lowering the rate of AOS production during stress. It may be used in its most prominent form as complete dormancy, similar to seed or winter bud dormancy, or it may be used in its moderate form, i.e., intermediate metabolic suppression, combined with the elevated expression of multiple defense pathways, to resist short-term exposures to stress.

3.8 Adaptive Plasticity as a Defense Strategy of Plants in the Field

The acclimation of plants to extreme habitats may require changes in plant morphology, physiology, development, growth pattern, reproductive timing or even offspring developmental patterns. Moreover, a single plant genotype may produce different anatomical and physiological phenotypes even within similar environments, a phenomenon known as phenotypic plasticity (Sultan 2000). The analysis of plants naturally growing within their ecosystem may therefore be very different from the study of plants in the laboratory, and large differences between seemingly similar plants can be expected. The analysis of *R. raetam* plants growing within their natural ecosystem revealed that different plants growing within the same habitat and subjected to similar growth conditions used different combinations of avoidance and resistance strategies. Thus, while some plants remained dormant even after the first rainfall of the season, other plants did not enter dormancy until very late during the dry season (cf. Fig. 3.2A, B). Interestingly, molecular analysis of the expression pattern of 12 different genes involved in AOS-removal, drought and heat stress response during the course of 12 months in *R. raetam* plants with different growth patterns revealed that the expression of some genes was primarily affected by the phenotypic variability (plasticity) of plants, whereas the expression pattern of other genes was primarily affected by environmental parameters and was therefore constant between different plants with different growth patterns (Merquiol et al. 2002). The large variability in the response of plants naturally growing within their ecosystem can be viewed as a type of defense strategy that will assure the survival of at least a few plants from the community during a catastrophic event, such as a stretch of years without rain. By what seems to be a "random" process, different plants in the collective may use a conserved growth pattern compared to others even during seasons rich in rain. In addition, not all plants will flower during a given season, and even when flowering, different plants will flower during different periods of the growth season. Unfortunately, the molecular mechanisms controlling these complex phenomena are poorly understood. Nevertheless, care should be taken to consider phenotypic plasticity when performing or analyzing data from a molecular study of gene expression in plants growing in nature.

3.9 Signalling Networks during Multiple Stress Conditions

The complexity of signalling events associated with a multiple stress condition is believed to involve calcium, calcium-regulated proteins, MAPK cascades and cross-talk between different transcription factors (Liu et al. 1998; Xiong et al. 1999; Bowler and Fluhr 2000; Kovtun et al. 2000; Knight and Knight 2001). Perhaps the earliest evidence for cross-talk between different stress-response signal transduction pathways originated from the phenomenon of "cross-tolerance", in which a specific stress condition induced in plants resistance to a subsequent, different, environmental stress (Bowler and Fluhr 2000). This suggested that different stresses may activate similar or overlapping pathways within the plant.

An enhanced rate of AOS production is believed to be a common denominator linking many of the different environmental stress conditions encountered by plants. Genes involved in AOS removal such as SOD or APX may therefore be an ideal substrate for studying how different signal transduction pathways activated during different stresses converge during a multiple stress situation. A good example of a study on stress signalling during multiple stresses comes from the work of Xiong et al. (1999, 2001a,b). Using a reporter gene (*luciferase*), fused to a promoter of a stress response gene (RD29A), they devised and undertook a series of genetic screens trying to identify common regulators that affect stress-response gene expression during drought, osmotic stress, low temperature, or exogenous abscisic acid (ABA) application, as well as a combination of osmotic stress, temperature and ABA. Their screens identified at least two regulators: a molybdenum cofactor sulfurase, which modulates cold and osmotic stress-response gene expression (Xiong et al. 2001a), and an inositol polyphosphate 1-phosphatase, a negative regulator of ABA and stress signalling in *Arabidopsis* (Xiong et al. 2001b). An alternative approach to study converging cellular pathways during multiple stresses is demonstrated by the work of Seki et al. (2001), which used a cDNA microarray screen to identify drought and cold responsive genes in a collection of clones obtained from cDNA libraries constructed from plants subjected to drought or cold stress. The screen of Seki and coworkers identified a number of genes that were induced by the two stresses and were also under the control of DREB1, a transcription factor that binds also to the RD29A promoter and whose expression in transgenic plants induces freezing tolerance (Jaglo-Ottosen et al. 1998).

Future work utilizing genomic wide arrays or chips screened with RNA samples obtained from plants subjected to different levels of environmental stresses, as well as to different combinations of stresses, may reveal some of the hierarchy and coordinated function of transcription factors involved in regulating gene expression during multiple stresses. A good example of chip analysis of hierarchic function of transcription factors originates from the work of Tepperman et al. (2001), who used Affymetrix chips to study the mode of action of phytochrome in *Arabidopsis*.

3.10 AOS at the Interface between Biotic and Abiotic Stresses, a Possible Conflict During Multiple Stresses

AOS play a central role in the defense response of plants against invading pathogens. During a plant pathogen interaction that results in the activation of plant defenses and suppression of pathogen infection, AOS are produced by plant cells at a very high rate. The production of AOS during this response is sometimes termed the "oxidative burst" and is thought to result from the facilitated enzymatic activity of NADPH oxidases, cell wall-bound peroxidases and amine oxidases in the apoplast. Hydrogen peroxide produced during this response is thought to diffuse into cells and activate many of the plant defenses, including the induction of programmed cell death (PCD; Hammond-Kosack and Jones 1996). The activity of key enzymes in the removal of H_2O_2 (APX and catalase) is suppressed during this response by the plant hormone salicylic acid and by nitric oxide that accumulate at least initially at and around the infection site (Clark et al. 2000; Klessig et al. 2000). It was also reported that during pathogen-induced PCD the expression of APX is post-transcriptionally suppressed (Mittler et al. 1998) and the expression of catalase is down-regulated at the steady-state mRNA level (Dorey et al. 1998). These findings point to a response in which the plant simultaneously produces more AOS and at the same time diminishes its own capability to scavenge H_2O_2, therefore resulting in an uncontrollable accumulation of AOS possibly to the point of cell death.

The role AOS play during the defense response of plants against biotic stresses appears to be contrary to the role AOS play during abiotic stresses in which the plant activates AOS-scavenging mechanisms in order to attenuate the rate of AOS production. This apparent conflict raises the question of how the plant will manipulate its own rate of AOS production when it comes under biotic attack during abiotic stress. One example that suggests the possibility of a conflict comes from comparing the rate of pathogen-induced PCD between control plants and plants that were previously subjected to oxidative stress (Mittler et al. 1999). It was found that plants that were previously subjected to oxidative stress and had a higher expression level of antioxidative enzymes had a reduced rate of PCD compared to unstressed control plants (Mittler et al. 1999). Thus, it appears as if the elevated level of AOS-scavenging enzymes that resulted from environmental stress interfered with the activation of PCD that required the elevated production of AOS. Further studies are however required to elucidate the complex role that AOS may play during the interaction between biotic and abiotic stresses. One possibility is that the activity of certain AOS-removal enzymes is controlled at different levels during biotic and abiotic stresses. For example, the steady-state transcript level of APX is induced during biotic and abiotic stresses; however, during a biotic stress another layer of regulation is activated and the expression of APX is suppressed post-transcriptionally (Mittler et al. 1998). Thus, the expression level of this key H_2O_2-removal enzyme is up-regulated at the level of the steady-state mRNA level in response to biotic and abiotic stresses, but the response to biotic stress is attenuated at the post-transcriptional level.

A constant conflict between biotic insults and abiotic stresses appears to exist in nature, and a number of studies suggest a link between the successful infection of plants and environmental stresses (Sandermann et al. 1997). Thus, in certain ecological habitats, abiotically stressed plants may be more sensitive to pathogen attack than unstressed plants. Future studies using advanced tools such as array technology may shed more light on this complex interaction between biotic and abiotic stresses.

References

Allan AL, Fluhr R (1997) Two distinct sources of elicited reactive oxygen species in tobacco epidermis cells. Plant Cell 9:1559–1572

Allen R (1995) Dissection of oxidative stress tolerance using transgenic plants. Plant Physiol 107:1049–1054

Asada K (1999) The water-water cycle in chloroplasts: scavenging of active oxygen and dissipation of excess photons. Annu Rev Plant Physiol Plant Mol Biol 50:601–639

Asada K (2000) The water-water cycle as alternative photon and electron sinks. Philos Trans R Soc Lond B 355:1419–1431

Asada K, Takahashi M (1987) Production and scavenging of active oxygen in photosynthesis. In: Kyle DJ, Osmond CB, Arntzen CJ (eds) Photoinhibition. Elsevier, Amsterdam, pp 227–287

Baier M, Dietz K-J (1999) Alkyl hydroperoxide reductases: the way out of the oxidative breakdown of lipids in chloroplasts. Trends Plant Sci 4:166–168

Bolwell GP, Wojitaszek P (1997) Mechanisms for the generation of reactive oxygen species in plant defense: a broad perspective. Physiol Mol Plant Pathol 51:347–366

Boveris A, Cadenas E (1982) Production of superoxide radicals and hydrogen peroxide in mitochondria. In: Oberley LW (ed) Superoxide dismutase, vol II. CRC Press, Boca Raton, pp 15–30

Bowler C, Fluhr R (2000) The role of calcium and activated oxygens as signals for controlling cross-tolerance. Trends Plant Sci 5:241–246

Bowler C, Van Montagu M, Inze D (1992) Superoxide dismutases and stress tolerance. Annu Rev Plant Physiol Plant Mol Biol 43:83–116

Chamnongpol S, Willekens H, Langebartels C, Van Montagu M, Inze D, Van Camp W (1996) Transgenic tobacco with reduced catalase activity develops necrotic lesions and induces pathogenesis-related expression under high light. Plant J 10:491–503

Clark D, Durner J, Navarre DA, Klessig DF (2000) Nitric oxide inhibition of tobacco catalase and ascorbate peroxidase. Mol Plant Microbe Interact 13:1380–1384

Dat J, Vandenabeele S, Vranova E, Van Montagu M, Inze D, Van Breusegem F (2000) Dual action of the active oxygen species during plant stress responses. Cell Mol Life Sci 57:779–795

Davidson JF, Schiestl RH (2001) Mitochondrial respiratory electron carriers are involved in oxidative stress during heat stress in *Saccharomyces cerevisiae*. Mol Cell Biol 21:8483–8489

Dorey S, Baillieul F, Saindrenan P, Fritig B, Kauffmann S (1998) Tobacco class I and class II catalases are differentially expressed during elicitor-induced hypersensitive cell death and localized acquired resistance. Mol Plant Microb Inter 11:1102–1109

Eshdat Y, Holland D, Faltin Z, Ben-Hayyim G (1997) Plant glutathione peroxidases. Physiol Plant 100:234–240

Foyer C, Halliwell B (1976) The presence of glutathione and glutathione reductase in chloroplasts: a proposed role in ascorbic acid metabolism. Planta 133:21–25

Foyer CH, Noctor G (2000) Oxygen processing in photosynthesis: regulation and signaling. New Phytol 146:359–388

Grant JJ, Loake GJ (2000) Role of reactive oxygen intermediates and cognate redox signaling in disease resistance. Plant Physiol 124:21–29

Groom QJ, Torres MA, Fordham-Skelton AP, Hammond-Kosack KE, Robinson NJ, Jones JDG (1996) Rboha a rice homologue of the mammalian gp91[phox] respiratory burst oxidase gene. Plant J 10:515–522

Halliwell B, Gutteridge JMC (1989) Free radicals in biology and medicine. Clarendon, Oxford

Hammond-Kosack KE, Jones JDG (1996) Resistance gene-dependent plant defense responses. Plant Cell 8:1773–1791

Jaglo-Ottosen KR, Gilmour SJ, Zarka DG, Schabenberger O, Thomashow MF (1998) *Arabidopsis* CBF1 overexpression induces COR genes and enhances freezing tolerance. Science 280:104–106

Karpinski S, Escobar C, Karpinska B, Creissen G, Mullineaux PM (1997) Photosynthetic electron transport regulates the expression of cytosolic ascorbate peroxidase genes in *Arabidopsis* during excess light stress. Plant Cell 9:627–640

Karpinski S, Reynolds H, Karpinska B, Wingsle G, Creissen G, Mullineaux P (1999) Systemic signaling and acclimation in response to excess excitation energy in *Arabidopsis*. Science 284:654–657

Keller T, Damude HG, Werner D, Doerner P, Dixon RA, Lamb C (1998) A plant homologue of the neutrophil NADPH oxidase gp91[phox] subunit gene encodes a plasma membrane protein with Ca^{2+} binding motifs. Plant Cell 10:255–266

Kendall AC, Keys AJ, Turner JC, Lea PJ, Miflin BJ (1983) The isolation and characterization of a catalase-deficient mutant of barley. Planta 159:505–511

Kessler A, Baldwin IT (2002) Plant responses to insect herbivory: the emerging molecular analysis. Annu Rev Plant Biol 53:299–328

Klessig DF, Durner J, Noad R, Navarre DA, Wendehenne D, Kumar D, Zhou JM, Shah J, Zhang S, Kachroo P, Trifa Y, Pontier D, Lam E, Silva H (2000) Nitric oxide and salicylic acid signaling in plant defense. Proc Natl Acad Sci USA 97:8849–8855

Knight H, Knight MR (2001) Abiotic stress signaling pathways: specificity and cross-talk. Trends Plant Sci 6:262–267

Kovtun Y, Chiu WL, Tena G, Sheen J (2000) Functional analysis of oxidative stress-activated mitogen-activated protein kinase cascade in plants. Proc Natl Acad Sci USA 97:2940–2945

Krause GH, Cornic G (1987) CO_2 and O_2 interactions in photoinhibition. In: Kyle DJ, Osmond CB, Arntzen CJ (eds) Photoinhibition. Elsevier, Amsterdam, pp 169–196

Langebartels C, Schraudner M, Heller W, Ernest D, Sandermann H (2000) Oxidative stress and defense reactions in plants exposed to air pollutants and UV-B radiation. In: Inze D, van Montagu M (eds) Oxidative stress in plants. Harwood Acad Publisher, London, pp 105–135

Liu Q, Kasuga M, Sakuma Y, Abe H, Miura S, Yamaguchi-Shinozaki K, Shinozaki K (1998) Two transcription factors, DREB1 and DREB2, with an EREBP/AP2 DNA binding domain separate two cellular signal transduction pathways in drought- and low-temperature-responsive gene expression, respectively, in *Arabidopsis*. Plant Cell 10:1391–1406

Maxwell DP, Wang Y, McIntosh L (1999) The alternate oxidase lowers mitochondrial reactive oxygen production in plant cells. Proc Natl Acad Sci USA 96:8271–8276

Merquiol E, Pnueli L, Cohen M, Simovitch M, Goloubinoff P, Kaplan A, Mittler R (2002) Seasonal and diurnal variations in gene expression in the desert legume *Retama raetam*. Plant Cell Environ 25:1627–1638

Mittler R (2002) Oxidative stress, antioxidants, and stress tolerance. Trends Plant Sci 7:405–410

Mittler R, Zilinskas BA (1994) Regulation of pea cytosolic ascorbate peroxidase and other antioxidant enzymes during the progression of drought stress and following recovery from drought. Plant J 5:397–405

Mittler R, Feng X, Cohen M (1998) Post-transcriptional suppression of cytosolic ascorbate peroxidase expression during pathogen-induced programmed cell death in tobacco. Plant Cell 10:461–474

Mittler R, Hallak-Herr E, Orvar BL, Van Camp W, Willekens H, Inze D, Ellis B (1999) Transgenic tobacco plants with reduced capability to detoxify reactive oxygen intermediates are hyper-responsive to pathogen infection. Proc Natl Acad Sci USA 96:14165–14170

Mittler R, Merquiol E, Hallak-Herr E, Rachmilevitch S, Kaplan A, Cohen M (2001) Living under a 'dormant' canopy: a molecular acclimation mechanism of the desert plant *Retama raetam*. Plant J 25:407–416

Mullineaux P, Karpinski S (2002) Signal transduction in response to excess light: getting out of the chloroplast. Curr Opin Plant Biol 5:43–48

Nakano Y, Asada K (1981) Hydrogen peroxide is scavenged by ascorbate-specific peroxidase in spinach chloroplasts. Plant Cell Physiol 22:867–880

Niyogi KK (1999) Photoprotection revisited: genetic and molecular approaches. Annu Rev Plant Physiol Plant Mol Biol 50:333–359

Niyogi KK (2000) Safety valves for photosynthesis. Curr Opin Plant Biol 3:455–460

Noctor G, Foyer C (1998) Ascorbate and glutathione: keeping active oxygen under control. Annu Rev Plant Physiol Plant Mol Biol 49:249–279

Orvar BL, Ellis BE (1997) Transgenic tobacco plants expressing antisense RNA for cytosolic ascorbate peroxidase show increased susceptibility to ozone injury. Plant J 11:1297–1305

Powles SB (1984) Photoinhibition of photosynthesis induced by visible light. Annu Rev Plant Physiol 35:15–44

Purvis AC, Shewfelt RL (1993) Does the alternate pathway ameliorate chilling injury in sensitive plant tissues? Physiol Plant 88:712–718

Rao MV, Davis K (1999) Ozone-induced cell death occurs via two distinct mechanisms in *Arabidopsis*: the role of salicylic acid. Plant J 17:603–614

Rizhsky L, Hallak-Herr E, Van Breusegem F, Rachmilevitch S, Rodermel S, Inzé D, Mittler R (2002a) Double antisense plants with suppressed expression of ascorbate peroxidase and catalase are less sensitive to oxidative stress than single antisense plants with suppressed expression of ascorbate peroxidase or catalase. Plant J 32:329–342

Rizhsky L, Hongjian L, Mittler R (2002b) The combined effect of drought stress and heat shock on gene expression in tobacco. Plant Physiol 130:1143–1151

Sandermann H, Wellburn AR, Heath RL (eds) (1997) Forest decline and ozone. Springer, Berlin Heidelberg New York

Sandermann H, Ernst D, Heller W, Langbartels C (1998) Ozone: an abiotic elicitor of plant defence reactions. Trends Plant Sci 3:47–50

Scandalios JG (1997) Molecular genetics of superoxide dismutases in plants. In: JG (ed) Oxidative stress and the molecular biology of antioxidant defenses. Cold Spring Harbor Laboratory Press, Cold Spring Harbor, NY, pp 527–568

Seki M, Narusaka M, Abe H, Kasuga M, Yamaguchi-Shinozaki K, Carninci P, Hayashizaki Y, Shinozaki K (2001) Monitoring the expression pattern of 1300 *Arabidopsis* genes under drought and cold stresses by using a full-length cDNA microarray. Plant Cell 13:61–72

Smirnoff N (2000) Ascorbate biosynthesis and function in photoprotection. Philos Trans R Soc Lond B 355:1455–1464

Stevens RG, Creissen GP, Mullineaux PM (1997) Cloning and characterisation of a cytosolic glutathione reductase cDNA from pea (*Pisum sativum* L.) and its expression in response to stress. Plant Mol Biol 35:641–654

Sultan SE (2000) Phenotypic plasticity for plant development, function and life history. Trends Plant Sci 5:537–542

Surplus SL, Jordan BR, Murphy AM, Carr JP, Thomas B, Mackerness SA-H (1998) Ultra-violet-B-induced responses in *Arabidopsis thaliana*: role of salicylic acid and reactive oxygen species in the regulation of transcripts encoding photosynthetic and acid pathogenesis-related proteins. Plant Cell Environ 21:685–694

Tepperman JM, Zhu T, Chang HS, Wang X, Quail PH (2001) Multiple transcription-factor genes are early targets of phytochrome A signaling. Proc Natl Acad Sci USA 98:9437–9442

Willekens H, Inze D, Van Montagu M, Van Camp W (1995) Catalases in plants. Mol Breed 1:207–228

Willekens H, Chamnongpol S, Davey M, Schraudner M, Langebartels C, Van Montagu M, Inze D, Van Camp W (1997) Catalase is a sink for H_2O_2 and is indispensable for stress defence in C-3 plants. EMBO J 16:4806–4816

Wingler A, Len PJ, Quick WP, Leegood RC (2000) Photorespiration: metabolic pathways and their role in stress protection. Philos Trans R Soc Lond B 355:1517–1529

Xiong L, Ishitani M, Zhu JK (1999) Interaction of osmotic stress, temperature, and abscisic acid in the regulation of gene expression in *Arabidopsis*. Plant Physiol 119:205–212

Xiong L, Ishitani M, Lee H, Zhu JK (2001a) The *Arabidopsis los5/aba3* locus encodes a molybdenum cofactor sulfurase and modulates cold stress- and osmotic stress-responsive gene expression. Plant Cell 13:2063–2083

Xiong L, Lee BH, Ishitani M, Lee H, Zhang C, Zhu JK (2001b) FIERY1 encoding an inositol polyphosphate 1-phosphatase is a negative regulator of abscisic acid and stress signaling in *Arabidopsis*. Genes Dev 15:1971–1984

4 Ethylene and Jasmonate as Regulators of Cell Death in Disease Resistance

CHRISTIAN LANGEBARTELS AND JAAKKO KANGASJÄRVI

4.1 Introduction: Input Signals and Plant Responses in Ecotoxicology

Plants in ecosystems are exposed in parallel to a variety of environmental challenges such as xenobiotics, heavy metals and air pollutants as well as biotic invaders such as microbial pathogens and herbivorous insects. They have developed effective recognition, signal transduction and defense mechanisms to counteract these abiotic and biotic stressors (Inzé and Van Montagu 1995; Karban and Baldwin 1997). These strategies include local responses at the site of insult as well as systemic responses throughout the affected organ and the entire plant. Moreover, communication through the gas phase and the root-soil interphase, the rhizosphere, is affected, which will most probably influence other organisms in the ecosystem (Van Loon et al. 1998; Farmer 2001). During the past years, evidence has accumulated showing that most of the abiotic and biotic stressors produce sublethal effects on field-grown plants, which nevertheless can have a greater effect on population size than does acute toxicity.

Ecotoxicological effects of pollutants result primarily from the actions and interactions between the physico-chemical characteristics of the biotopes, the structural and functional properties of the living organisms and the contamination modalities. Ecotoxicological mechanisms primarily depend on the bioavailability of the toxic products in an ecosystem (Moriarty 1988). Numerous physical, chemical and biological processes control the chemical fate of contaminants. In addition, accessibility to the biological barriers that separate the organisms from the surrounding medium depends directly on bioavailability. In recent years, advances have been made in the understanding of detrimental effects of various compounds, as well as of signal transduction, innate immunity responses and individual plant fitness (Sandermann 1994; Cohn et al. 2001; Heil and Baldwin 2002). This new knowledge should lead to improvements in monitoring plant health and predicting the impact of toxi-

Ecological Studies, Vol. 170
H. Sandermann (Ed.)
Molecular Ecotoxicology of Plants
© Springer-Verlag Berlin Heidelberg 2004

cants on plant populations, which is a fundamental ecotoxicological goal. Even slight, non-significant influences on single components within regulatory cascades, like cellular division or signal transduction, that would not result in any discernible effect, might ultimately, through sequential propagation or through interaction with additional, unrelated factors, affect a whole population by its negative consequences on fitness: disturbances in hormonal homeostasis, immunological status as well as signal transduction and gene activation. Therefore, a more "mechanism-based" approach to the experimental investigation of potential hazards through the contamination of terrestrial ecosystems by abiotic and biotic stress factors should yield more meaningful results and insights than the normally used standard battery of ecotoxicology assays (Seiler 2002).

4.2 Cell Death and Disease Resistance in Plants

Evolutionary conserved targets and basic mechanisms of cellular function play a pivotal role in the context of ecotoxicology. Signal transduction and cell death mechanisms are generally well conserved in evolution and can thus be identified throughout the living world from unicellular to mammalian organisms (Dangl and Jones 2001; Lam et al. 2001). The response of plants to avirulent pathogens and non-pathogens was termed "hypersensitiveness" by Stakman (1915) to describe rapid cell death of leaf tissue at and around the invasion site. Later, the term 'hypersensitive response' (HR) was introduced, which comprises rapid, localized death of plant cells in association with the functional suppression of pathogen growth by the accumulation of defense-related metabolites and proteins at the infection site (Goodman and Novacky 1994). It is now widely accepted that this general defense strategy of plants is a form of genetically defined, programmed cell death (pcd; Pennell and Lamb 1997; Heath 2000; Mittler and Rhizhky 2000), with analogies to animal cell death, and that it is "hypersensitive" only with respect to the highly efficient perception and guard mechanisms that constantly survey the environment for potential aggressors.

Although the HR is a common feature of many resistance reactions, it is not an obligatory component. Several reactions such as those mediated by the *mlo* gene in barley and *Cf* genes of tomato under high humidity precede the induction of HR symptoms or occur without visible HR, respectively (Shirasu and Schulze-Lefert 2000). HR lesions are also formed spontaneously in the absence of pathogens in various mutants (Mittler and Rizhsky 2000), which indicates that it is under a pre-existing genetic control. In addition to the well-studied HR, pcd is observed during symptom formation following virulent infections (Morel and Dangl 1997), is induced by the atmospheric pollutant, ozone (Overmyer et al. 2002), and is involved in various developmental processes such as deletion of suspensor and aleurone cells, formation of leaf lobes and perforations and cell death in xylem tracheary elements (Pennell and Lamb 1997; Fukuda 2000).

Necrotic death has to be differentiated from programmed cell death. Necrosis is accidental cell death by overwhelming physical or chemical trauma. It is disorganized and uncontrolled, and usually results in the death of large groups of cells or whole tissues. Necrotic cells lose membrane and organelle integrity early, and DNA fragmentation, if it occurs, is random and appears late. The key characteristics of pcd are the cell's autonomous decision to die (triggered from within) and active participation of the cell in its own execution. Apoptosis in animal cells is a form of pcd involving a series of well-organized events requiring active cell participation. It is the basis for normal tissue remodeling as well as the end result of toxic insults. The key features of apoptosis are nuclear shrinkage, DNA fragmentation, late loss of organelles, cell shrinkage and break up of the cell and nucleus into membrane-bound apoptotic bodies, which are taken up by the surrounding cells (McConkey and Orrenius 1994). At the molecular level, the fragmentation of nuclear DNA into approximately 50-kb pieces (oligonucleosomes) is involved in this cell disassembly. Several genes and proteins including fas, the bcl family, caspases and other proteases, and the release of cytochrome c from mitochondria into the cytosol have been identified as regulators of pcd in different species (Kidd et al. 2000).

Findings in recent years point to similarities in the processes involved in plant and animal pcd. However, because of the physical restriction of the cell wall, plant cells can obviously never fulfill all the criteria that define apoptosis in animal cells. Thus, the more general term "programmed cell death" is often favored in plants instead of "apoptosis". Caspase-like protease activity is involved in the N-gene mediated HR in tobacco (Del Pozo and Lam 1998), in the ozone-induced cell death in *Arabidopsis thaliana* (Overmyer et al. 2002) and cell death in barley cultures (Korthout et al. 2000). In addition, homologues of metacaspases, caspase-related proteases and of Bcl2 and related death-agonist proteins have been identified in plants (Dion et al. 1997; Uren et al. 2000). Regarding signal transduction, various structurally conserved mitogen-activated protein kinases are present in yeast, plant and animal cells (Meskiene and Hirt 2000). Finally, a role for reactive oxygen species (ROS) following various initiating stimuli has been supported in animals (Robertson and Orrenius 2000) and plants (Grant and Loake 2000; Scheel 2002).

In addition to the similarities between animal and plant cell death, there are also plant-specific regulatory systems, plant hormones, which are involved in the control of cell death and defense responses. In this chapter, we will focus on ethylene and jasmonate, two plant-derived signal compounds that are key players in plant cell death and disease resistance (Van Camp et al. 1998; Overmyer et al. 2000; Langebartels et al. 2002a,b). Their involvement as regulators of cell death will be described, aside from their role in plant-pathogen interactions, utilizing a model system that has been used to unravel ROS-mediated cell death in plants: ozone-triggered cell death in sensitive plants (Sandermann et al. 1998; Kangasjärvi et al. 2001; Rao and Davis 2001) We will review recent results on cooperating effects of ethylene and jasmonate in regulating the expression of defense genes and their potential antagonistic roles in cell death processes. We also describe the emerging picture that defense genes and cell death in plants are regulated by a small number of signal pathways also including those for salicylic acid, NO and H_2O_2. Cross-talk between these pathways appears to be very common and important. Other roles of ethylene and jasmonate, e.g., in plant development, have been described in the excellent reviews by Hamberg and Gardner (1992), Brown (1997), Lelièvre et al. (1997), Morgan and Drew (1997), Johnson and Ecker (1998) and Wang et al (2002). In the context of ecotoxicology, we discuss that both ethylene and jasmonate are effective as growth and defense regulators in the gaseous state, thereby potentially signalling stress, developmental and toxic effects from cell to cell, individual to individual, from one species to another and even leading to interkingdom signal interference.

4.3 Biosynthesis and Perception of Ethylene in Plants

4.3.1 Mechanism of Ethylene Biosynthesis

There are several ways in which the gaseous hormone ethylene (ethene) influences plant growth and development as well as plant health and fitness. It regulates many developmental processes such as seed germination, tissue differentiation, flowering initiation, flower opening and senescence, fruit ripening and degreening, production of volatile organic compounds involved in aroma formation and defense, and leaf and fruit abscission (Fig. 4.1; Abeles et al. 1992; Brown 1997; Lelièvre et al. 1997; Llop-Tous et al. 2000). In addition, ethylene is an important mediator of plant responses to biotic and abiotic stresses (Kende 1993; Pieterse and Van Loon 1999; Wang et al. 2002). Ethylene has mainly been implicated in the response to virulent pathogens, but also to environmental challenges like the air pollutant ozone (Langebartels et al. 2002a), xenobiotics as well as temperature extremes and drought (Morgan and Drew 1997). It is produced in all higher plants as well as by a variety of microorganisms including important plant pathogens such as *Pseudomonas syringae*. It is also formed non-enzymatically during decomposition of various materials. Many organisms have perception and response systems for ethylene, even though the controlled production via the "ethylene forming enzyme" ACC oxidase seems to be restricted to gymnosperms and angiosperms (John 1997).

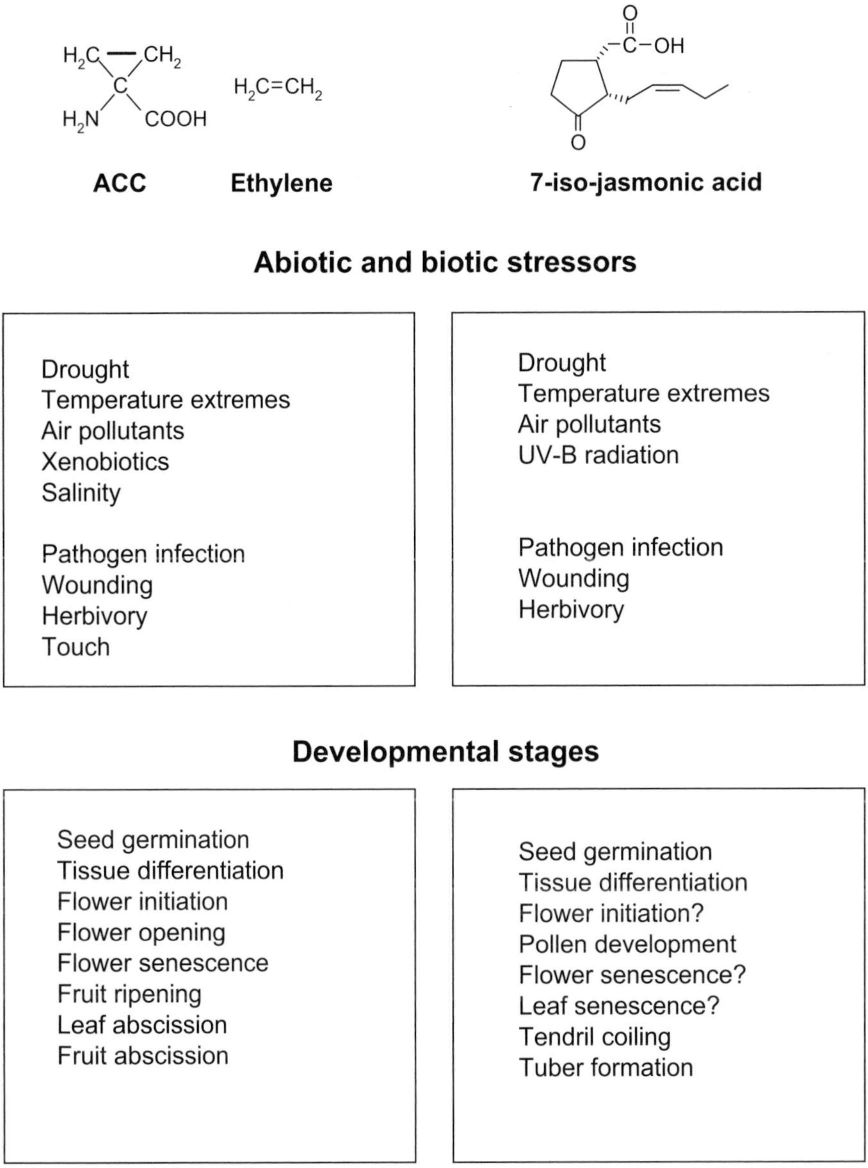

ACC   **Ethylene**   **7-iso-jasmonic acid**

Abiotic and biotic stressors

Drought Temperature extremes Air pollutants Xenobiotics Salinity Pathogen infection Wounding Herbivory Touch	Drought Temperature extremes Air pollutants UV-B radiation Pathogen infection Wounding Herbivory

Developmental stages

Seed germination Tissue differentiation Flower initiation Flower opening Flower senescence Fruit ripening Leaf abscission Fruit abscission	Seed germination Tissue differentiation Flower initiation? Pollen development Flower senescence? Leaf senescence? Tendril coiling Tuber formation

Fig. 4.1. Environmental stress and developmental stages of plants that involve ethylene and/or jasmonate action. *ACC*, 1-Aminocyclopropane-1-carboxylic acid, the direct precursor of ethylene in higher plants. (Modified from Abeles et al. 1992; Reymond and Farmer 1998)

    While the main steps in ethylene biosynthesis of higher plants were unraveled between 1970 and 1990 (Yang and Hoffman 1984; Abeles et al. 1992), key components of its signal transduction pathway, and, more recently, also the complex regulation of ethylene biosynthesis are a current topic of intense research (Stepanova and Ecker 2000; Wang et al. 2002). Together, the ethylene biosynthesis, perception and transduction system represents one of the most detailed pathways known for hormonal action in plants and animals. The important methyl donor and polyamine as well as ethylene precursor S-adenosyl methionine (SAM) is produced by SAM synthetase (E.C. 2.5.1.6.) The first committed step in ethylene biosynthesis, the conversion of S-adenosyl methionine to 1-aminocyclopropane-1-carboxylic acid (ACC), is catalyzed by ACC synthase (ACS, E.C. 4.4.1.14; Fig. 4.1). ACC oxidase (ACO), in turn, is the 'ethylene-forming enzyme' and oxidizes ACC to ethylene, CO_2 and cyanide (Abeles et al. 1992). Both ACS and ACO are encoded by gene families consisting of at least eight (Oetiker et al. 1997; Shiu et al. 1998) and four members (Barry et al. 1996; Nakatsuka et al. 1998) in tomato, respectively. In *A. thaliana*, seven ACS genes have been characterized experimentally, but in the complete genomic sequence of *Arabidopsis* at least 12 ASC and four ACO genes have been predicted.

    For a long time ACC synthase has been regarded as the rate-limiting step in ethylene biosynthesis (Kende 1993). The ACS genes show differential induction by wounding, ozone exposure, pathogen attack and elicitor treatment and both positive and negative feedback regulation at the transcriptional level has been described in different systems (Lincoln et al. 1993; Liu et al. 1993; Oetiker et al. 1997; Barry et al. 2000; Moeder et al. 2002). In addition, ACS is known to be regulated posttranscriptionally by phosphorylation (Felix et al. 1991; Spanu et al. 1994). For example, the tomato ACS2 is phosphorylated at the C-terminal conserved serine-460 (Tatsuki and Mori 2001). The ethylene-overproducing *Arabidopsis* mutant *eto2-1* has a single base insertion upstream of the codon for this conserved serine in *At-ACS5*, which thus changes the reading frame in the C-terminus of *At-ACS5* and results in an enzyme that cannot be phosphorylated. Correspondingly, ethylene production in the etiolated seedlings of the *eto2-1* mutant is nearly 20-fold higher than in wild-type plants (Vogel et al. 1998). More on the possible molecular mechanisms involved in ACS regulation can be found in Wang et al. (2002).

    ACC oxidase belongs to the family of Fe(II)-dependent dioxygenases. Unlike other members of this family, it uses ascorbate instead of 2-oxoglutarate as a co-substrate and depends on CO_2 as co-factor. ACC oxidase has been believed to be constitutively present in plant cells (Kende 1993). Recent results, however, show that distinct isoforms of ACO are stimulated during certain developmental stages and by external stimuli such as wounding, pathogen infection, flooding and ozone exposure (Avni et al. 1994; Knoester et al. 1995; Tuomainen et al. 1997; Moeder et al. 2002).

4.3.2 Ethylene Perception by Two-Component Systems, Sensors for Environmental Stress

Ethylene signal transduction starts with the ETR (ethylene-resistant) family of ethylene receptors, of which five members are found in *Arabidopsis* and six in tomato (Wang et al. 2002). The prototype ETR1 is a modular protein with three trans-membrane domains that contain the ethylene-binding site as well as conserved histidine kinase and receiver domains. The latter two exhibit significant homology to bacterial two-component systems (TCS), which in their simplest form are composed of a histidine kinase and a response regulator (receiver and output domain; Parkinson and Kofoid 1992; Lohrmann and Harter 2002). The TCS sensor proteins in the periplasmic space of bacteria continuously monitor the surrounding medium for distinct environmental stimuli such as nutrients (C, P, N), toxic products (heavy metals, xenobiotics), osmolarity and redox status (Parkinson and Kofoid 1992). In addition, they are involved in light sensing as well as in the interactions of *Rhizobium* and *Agrobacterium* species with their host plants and in quorum sensing in *Erwinia* in the beginning of the pathogenic phase (Pierson et al. 1998). In the "hybrid kinase"-type of TCS both the histidine kinase and the receiver domain have fused together in one protein (Fig. 4.2; Parkinson and Kofoid 1992; Sakakibara et al. 2000). More than 45 homologues of bacterial TCS proteins (sensor kinases, response regulators and phosphotransfer proteins) have been identified in *A. thaliana* up to now (D'Agostino and Kieber 1999; Lohrmann and Harter 2002). They function in ethylene and cytokinin signalling, both of which comprise sensor kinases, but are also involved in other pathways such as light signalling in interaction with phytochrome (Kakimoto 1998; Sakakibara et al. 1999, 2000; Lohrmann and Harter 2002). Regarding the need of plant cells to monitor and integrate a multitude of inputs (e.g., in the composition of the surrounding apoplastic fluid), it is highly probable that TCS in plants are involved in the perception of additional inputs from abiotic and biotic stressors.

All five *Arabidopsis* ethylene receptors, ETR1, ERS1, ETR2, EIN4 and ERS2, share the N-terminal ethylene-binding domain (Ecker 1995). Structurally, the receptors form two classes: ETR1, ETR2 and EIN4 represent the hybrid-kinase type TCS and have receiver domains at their C terminus when ERS1 and ERS2 lack the receiver domain (Lohrmann and Harter 2002; Wang et al. 2002). However, these classes do not reflect the function of the receptors; they are classified into two functional groups based on the presence (subfamily I) or absence (subfamily II) of the canonical histidine kinase domain (Fig. 4.2). Members of the subfamily I (ETR1 and ERS1) have three predicted transmembrane domains, possesses a well-conserved His-kinase domain and function as homodimers. The subfamily II members (ETR2, EIN4 and ERS2) have four transmembrane domains and a degenerate histidine kinase domain lacking necessary subdomains essential for the catalytic activity. The in vitro

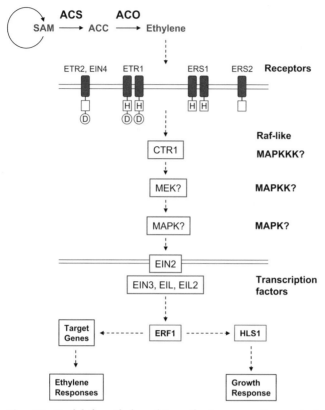

Fig. 4.2. Model for ethylene biosynthesis, perception and signal transduction. *ACC* Aminocyclopropane-1-carboxylic acid; *ACO* ACC oxidase; *ACS* ACC synthase; *CTR1* Raf-like kinase; *EIN2* Nramp metal transporter protein; *EIN3, EIL, EIL2* nuclear located transcription factors; *ERF1* transcription factor of the EREBP family; *ETR1, 2, EIN4, ERS1, 2* ethylene receptors with conserved histidine (H) and aspartate (D) residues involved in phosphorylation. *ETR2, EIN4* and *ERS* lack the canonical histidine kinase domain; *HLS1,* transcription factor involved in growth responses; *MAPK, MAPKK, MEK* postulated MAP kinase cascade involved in ethylene signalling; *SAM* S-adenosyl methionine

histidine autophosphorylation that has been demonstrated for ETR1 is not, however, required for the ethylene signal transmission (Wang et al. 2002). It has been postulated that the subfamily II, like bacterial proteins and phytochromes, may not function as histidine kinases, but have phosphatase or serine/threonine kinase activities. It has been shown in a heterologous yeast system that ETR1 is membrane-associated (Chang and Stadler 2001), putatively localized to the endoplasmic reticulum (Chen et al. 2002). Ethylene binding to ETR1 requires copper as co-factor, and the *RAN1* (responsive to antagonist) gene (Hirayama et al. 1999) in *A. thaliana*, a copper transporter, seems to deliver copper for functional ethylene receptors.

4.3.3 Ethylene Signal Transduction

While dominant mutations of ethylene receptors confer ethylene insensitivity, recessive loss-of-function mutations have wild-type phenotypes. When, however, three or more receptors are knocked out, constitutive ethylene responses are observed (Hua and Meyerowitz 1998). This suggests that in the absence of ethylene the receptors are negative regulators of the pathway. The downstream component CTR1 (constitutive triple response), which has similarity to RAF-like Ser/Thr kinases that initiate mitogen-activated protein kinase cascades in mammals, also negatively regulates ethylene responses by keeping a downstream component, the (membrane-localized) Nramp-type metal transporter EIN2 inactive (Fig. 4.2). However, it is not yet clear how CTR1 represses the downstream components of the pathway. Physical interaction between ETR1 and ERS1 on one side and CTR1 on the other has been described, forming a complex of these two as well as probably additional proteins (Clark et al 1998).

When ethylene binding inactivates the receptors, CTR1 is inactivated, thus relieving EIN2 from the repression by CTR1, and ethylene responses are activated. Genetic evidence indicates that EIN2 in turn is involved in the activation of (nuclear) EIN3-like transcription factors with at least six members in *A. thaliana* (Solano et al. 1998) and three in tomato (LeEIL1–3). The promoter of ERF1, an ethylene response element binding protein (EREBP; Fig. 4.2), contains a primary ethylene response element and has been postulated as target promotor for the EIN3 family (Solano and Ecker 1998). EREBPs, a plant-specific family of transcription factors, were initially identified through their binding to GCC boxes, promoter motifs associated with ethylene-induced gene expression, e.g., in PR proteins. The topic of gene regulation by *cis* elements and transcription factors will be treated in Chapter 6 (this Vol.).

Given the rapid activation and inactivation of the ethylene burst and the subsequent responses, it is surprising that ethylene binding to the receptor is quite stable with dissociation rates in the yeast-expressed ETR1 of more than 10 h (Bleecker 1999). This raises the question how can the ethylene response be down regulated so rapidly after the initial association of the hormone with the receptor. It is thought that the (ethylene-induced) synthesis of new, unoccupied receptor molecules in response to stress is responsible for the reactivation of CTR1 and thereby rapid elimination of the ethylene responses. Induction of certain receptor isoforms has been described after ethylene treatment (Wilkinson et al. 1995) and ozone stress in tomato (Moeder et al. 2002). In the latter case, ozone exposure caused the transcript levels of *Le-ETR1* (ETR1 class) to increase 1 h after the beginning of the exposure. Expression of the *NR* (*Le-ETR3*) receptor was unaffected by ozone, while the transcripts of *Le-ETR2* first decreased and then increased again (Moeder et al. 2002). There is much to be learned about this gene family in plants, but with the array of tools becoming available to *Arabidopsis* and tomato researchers, the pathways for ethylene signal transduction and TCS perception and transduction systems may soon reveal their remaining secrets.

4.3.4 Regulation of Ethylene Biosynthesis and Perception

Individual members of the ACS, ACO and ethylene receptor gene families are differentially expressed during developmental processes and in response to external stimuli (Rottmann et al. 1991; Lincoln et al. 1993; Barry et al. 1996, 2000; Oetiker et al. 1997; Tatsuki and Mori 1999; Ciardi et al. 2000). TMV infection, salicylic acid, ethylene and methyl jasmonate all induced *Ng-ACO1* and *Ng-ACO3*, but not *Ng-ACO2* in *Nicotiana glutinosa*. A recent study has shown that treatment of tomato plants with the air pollutant ozone could affect the expression of these gene families in parallel (Moeder et al. 2002). The changes in expression patterns were grouped into two classes based on temporal changes in expression. Rapid changes, within 1 h of the beginning of the treatment, were seen for *Le-ACS6, Le-ACO1, Le-ACO3, Le-ETR1* and *Le-ETR2*. Slower changes, occurring after 2 h, were detected for *Le-ACS2, Le-ACO2* and *Le-ACO4*. These results suggested a bi-phasic regulation of ethylene synthesis and perception in response to ozone. A similar biphasic pattern of ACC synthase induction was described as 'system 1' and 'system 2' ethylene production during tomato fruit ripening, which involves high levels of sustained ethylene production over many days (Barry et al. 1996, 2000). A biphasic relationship between the expression of *Le-ACS6* (rapid, transient response) and *Le-ACS2* (late accumulation) was also found in response to mechanical wounding of tomato leaves and mature green fruit (Tatsuki and Mori 1999).

The ethylene receptor gene family in tomato shows differential expression patterns in tissues and in response to stress factors (Lashbrook et al. 1998; Tieman and Klee 1999; Ciardi et al. 2000). Three family members, *Le-ETR1, Le-ETR2* and *NR*, were differentially regulated in response to ozone treatment (Moeder et al. 2002), while *NR* and *Le-ETR4* responded differentially to pathogen infection (Ciardi et al. 2000). The spatial location of ethylene biosynthesis in response to ozone has been analyzed with transgenic plants expressing a *Le-ACO1* promoter: GUS fusion (Fig. 4.3; Moeder et al. 2002). The results indicate that, rather than homogeneous expression throughout the leaf, GUS expression is confined to distinct regions surrounding the vascular tissue. Closer examination revealed that expression was mainly localized in the parenchyma cells. This restricted expression is of interest as clearly not all cells are responding to ozone in the same way. One hypothesis is that the bi-phasic pattern of *ACS, ACO* and *ETR* transcript accumulation in response to ozone treatment may be due to differential cellular localization of gene expression.

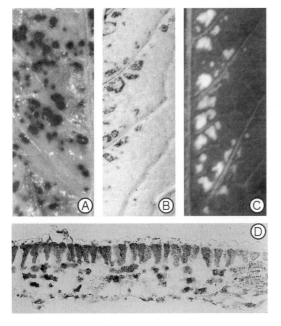

Fig. 4.3A–D. (in color at end of book) Colocalization of gene expression of the 'ethylene-forming enzyme' ACC oxidase, hydrogen peroxide accumulation and cell death in ozone-treated tomato plants (from Moeder et al. 2002). **A** GUS activity regulated by the Le-ACO1 promoter 1 h after the onset of ozone exposure. **B** H_2O_2 accumulation after 7 h. **C** Ozone-induced cell death after 24 h. **D** Transverse section through a leaflet showing GUS staining after 1 h

4.3.5 Ethylene Perception and Responses in Other Organisms

Ethylene is released during decomposition of various organic compounds, and it may have been available to early land plants as an indicator of stress. It has been speculated that its precursor, ACC, accumulated in land plants as a side product of S-adenosyl methionine metabolism, which is primarily used in various methylation reactions and in the biosynthesis of the ubiquitous polyamines spermidine and spermine (Galston et al. 1997). ACO activity was found in vivo and in vitro in angiosperms, and in vitro in seedlings of the Coniferales and Gnetales, but not of the Cycadaceae and Gingkoaceae (John 1997). It is thus assumed that ACO arose relatively late in the evolution of land plants. Ethylene as a gaseous hormone, however, can be perceived and responded to by the majority of plants as well as by certain bacteria and fungi, and may therefore act as a general "warning" signal across species in ecosystems.

4.4 Biosynthesis and Perception of Jasmonate in Plants

4.4.1 Biosynthesis of Signals of the Jasmonate Family

Jasmonate as the lead molecule of the jasmonate family of plant growth regulators is involved in the defense against insect parasites, possibly also to bacterial and fungal pathogens, as well as in wounding and desiccation responses (Fig. 4.1; Reymond and Farmer 1998; Howe and Schilmiller 2002; Liechti and Farmer 2002). Exogenous supply of jasmonate induces jasmonate-inducible proteins (JIPs) and various defense-related metabolites. Jasmonate may also enhance, in certain systems, senescence phenomena such as chlorophyll loss, lowering of photosynthesis and degradation of photosynthetic enzymes, e.g., Rubisco. In addition, jasmonate has been implicated as a long-distance endogenous wounding signal, activating, e.g., nicotine biosynthesis in the roots after leaf attack by herbivores (Baldwin et al. 2001; Kessler and Baldwin 2002).

Jasmonate and its methyl ester, methyl jasmonate (MeJa), are metabolites of polyunsaturated fatty acids, collectively called oxylipins (Fig. 4.4). Their biosynthesis is initiated by oxygenation of linoleic and linolenic acid by lipoxygenase (LOX) activity, a non-heme iron dioxygenase that introduces oxygen to either C_9 or C_{13} positions of linoleic and linolenic acid. Hydroperoxy products of LOX activity lead to a variety of oxylipin products, with two main routes for 13-hydroperoxides and three main routes for 9-hydroperoxides (Reymond and Farmer 1998; Howe and Schilmiller 2002). Jasmonate is formed via the allene oxide synthase (AOS) branch of the 13-LOX pathway, which leads from 13-hydroperoxy linolenic acid (13-HPOT) to the metabolic precursor 12-oxo-phytodienoic acid (12-OPDA), jasmonate and methyl jasmonate (Fig. 4.4). An alternative pathway, starting from C_{16} fatty acids, leads to the production of dinor-OPDA (Weber et al. 1997). In addition, various conjugates, e.g., amino acid conjugates of jasmonate and galactolipid conjugates of 12-OPDA, are formed.

The hydroperoxide lyase (HPL) branch of the 13-LOX pathway produces volatile C_6 aldehydes and alcohols as well as C_{12} keto fatty acids such as traumatic acid. C_6 aldehydes and alcohols make up the "green odor" of leaves and fruits and form a major part of the emission "bouquet" of plants under abiotic and biotic stress. They are currently analyzed for their roles in pathogen and insect defense (see below, part 9). Finally, divinyl ether synthase (DES) gives rise to divinyl ether fatty acids such as etherolenic acid, of which the functions are currently unknown. It is worth noting that all three key enzymes, AOS, HPL and DES, are closely related cytochrome P450 proteins with relative specificity for 9- or 13-hydroperoxides organizing the response into discrete 9-LOX and 13-LOX pathways (Howe et al. 2000; Howe and Schilmiller 2002).

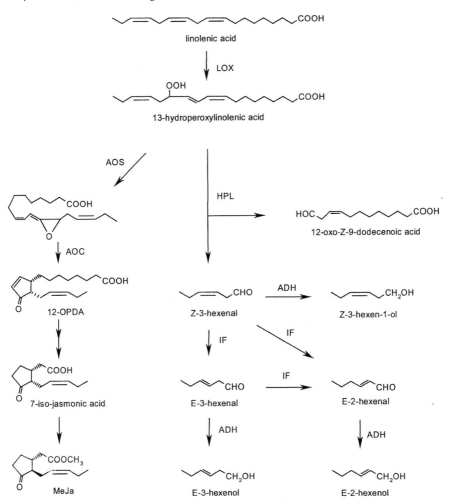

Fig. 4.4. Model for the biosynthesis of oxylipins and C6 volatiles in plants. *ADH* Alcohol dehydrogenase; *AOC* allene oxide cyclase; *AOS* allene oxide synthase; *HPL* hydroperoxide lyase; *IF* isomerization factor; *LOX* lipoxygenase; *MeJa* methyl jasmonate; *12-OPDA* 12-oxo-phytodienoic acid

Metabolism of 9-hydroperoxy fatty acids is also based on AOS, HPL and DES activities and leads to related oxylipins that perform an essential role in plant pathogen defense. Antisense plants for *9-LOX*, for example, showed enhanced susceptibility towards both avirulent and virulent fungal pathogens (Rancé et al. 1998) as well as fungal elicitors (Rustérucci et al. 1999). The AOS branch produces ketols and the "jasmonate-like" cyclopentenone 10-OPDA (Hamberg and Gardner 1992). HPL activity leads to products 9-oxo nonanoid acid and the volatile nondienal, which is also known in animal signal transduction. Increasing evidence has been obtained that the role of oxylipins in plants is comparable to that of the eicosanoid group of metabolites in animals.

4.4.2 Jasmonate Perception and Signalling

In a similar way as other hormones, the jasmonate signal is most probably first transduced via a receptor that binds the hormone. However, no receptor proteins that bind jasmonate have been identified so far. Exhaustive mutant screenings with *A. thaliana* have been conducted to identify and characterize the jasmonate receptor, but these screenings have predominantly revealed alleles of two jasmonate insensitive mutants *coi1* and *jar1*, neither one of which have been shown to act as receptor for jasmonate (Turner et al. 2002). In addition to *jar1* and *coi1*, several other jasmonate response mutants, such as *jin1* (Berger et al. 1996), which is insensitive to jasmonate and *cev1*, which has constitutive expression of JA-regulated genes such as *VSP1* (Ellis and Turner 2001), have been identified. The identification of genes encoding COI1 (Xie et al. 1998), JAR1 (Staswick et al. 2002) and CEV1 (Ellis et al. 2002) has revealed quite unrelated processes, which, when mutated, confer altered jasmonate responses in plants. Of these three, COI1, a F-box protein related to TIR, which is a part of the SCF[coi1] ubiquitin ligase complex (Xu et al. 2002), is most probably a genuine component of jasmonate signal transduction (Turner et al. 2002). However, the demonstrated functions of JAR1, an adenyl transferase that directly adenylates jasmonate, and CEV1, a cellulose synthase, demonstrate the complexity of jasmonate-related responses and the possible involvement of cell wall components.

Jasmonate signalling/action has two intriguing similarities with auxin since the molecular processes where both COI1 and JAR1 act have very close and well-studied counterparts in the auxin pathway. The recent cloning of *JAR1* has shown that JAR1 is most probably not directly involved in jasmonate signalling. JAR1 was shown to be closely similar to a soybean GH3, which is involved in auxin responses. In vitro, JAR1 can directly modify jasmonate by adenylating it (Staswick et al. 2002). Further molecular and substrate characterization of the 19 members of the *JAR1* gene family in *A. thaliana* showed, first of all, that JAR1 was the only member of the family that could directly adenylate jasmonate, and, secondly, that most other members of the family had adenylating activity on auxin. Furthermore, one of the family members was shown to adenylate salicylic acid. How this molecular function of JAR1 relates to the jasmonate insensitivity of the *jar1* mutant is currently not known. Furthermore, the significance of auxin and salicylic acid adenylation is also unresolved.

The COI1 protein belongs to the family of the F-box proteins involved in the E3 ubiquitin ligase complex known as the SRC complex (Xie et al. 1998). This complex is involved in the marking of specific proteins for degradation by the 26S proteasome complex by conjugation of ubiquitin to the target protein. Similar regulated protein targeting for degradation is well known in the auxin response, where ubiquitinylation by the SRC complex of proteins that repress the transcription of early auxin-responsive genes results in their transcriptional activation. The substrates of COI1 remain unidentified, but it has been speculated that they may be some of the key regulators of JA response (Turner et al. 2002).

4.5 Roles of Ethylene under Oxidative and Pathogen Stress

The role of ethylene in disease resistance is ambiguous and apparently depends on the plant-pathogen system studied: ethylene-insensitive mutants are more susceptible to the necrotrophs *Botrytis cinerea* and *Erwinia carotovora* suggesting that ethylene is a factor in basal resistance (Norman-Setterblad et al. 2000; Thomma et al. 2001). On the other hand, the ethylene-insensitive mutant *ein2* was relatively tolerant to infection with virulent strains of biotrophic *Pseudomonas* and *Xanthomonas* (Bent et al. 1992). As reported by Knoester et al. (1998), ethylene-insensitive tobacco was highly susceptible to typically non-pathogenic *Pythium* soil bacteria. These results suggest a promoting role for ethylene in regulating the spread of cell death in biotrophic, pertotrophic and non-host interactions. Following infection with necrotrophic pathogens, however, ethylene seems to inhibit uncontrolled cell death. The exact mechanisms of disease lesion development and the possible involvement of other signal molecules and programmed cell death therein are currently being resolved (see below).

Ethylene emission is one of the fastest plant responses to ozone, and it has been shown to correlate to ozone sensitivity in various species (Tingey et al. 1976). Initially, it was proposed that ozone could react chemically with ethylene and form radicals that in turn would damage the biological structures of the cells (Elstner et al. 1985; Mehlhorn and Wellburn 1987). More recent results, however, suggest that ethylene may play a more active role in cell death related to ozone damage. In *Arabidopsis* mutants selected for increased sensitivity to ozone (Overmyer et al. 2000), highly enhanced ethylene evolution was triggered during early lesion development when the Col-0 wild type showed only a minor response. Inhibitors of ethylene biosynthesis and perception reduced the accumulation of ROS and cell death in both tomato and *Arabidopsis* (Tuomainen et al. 1997; Overmyer et al. 2000; Moeder et al. 2002). Thus, both existing sensitivity of the genotype to ozone (tomato) and gain of sensitivity by mutation in resistant background (*A. thaliana*) involve rapid activation of ethylene biosynthesis in the tissues that subsequently exhibit cell death and visible symptom formation. It has been shown that the ethylene-dependent, ozone-induced cell death and visible lesion formation requires transcription, translation and specific protease action, and also otherwise fulfills the definitions of pcd (Overmyer et al. 2002). Thus, as has been predicted in several recent reviews (Sandermann et al. 1998; Rao et al. 2000; Kangasjärvi et al. 2001; Langebartels et al. 2002a), ozone seems indeed to intercept the preexisting cell death regulation by degradation in the cell wall to reactive oxygen species that are misinterpreted by the cells as cell death signals and accidentally misfire the cell death program. Furthermore, ethylene seems to be intimately involved in the regulation of active ROS production in the plant and thus is directly involved in the regulation of the magnitude and extent of cell death.

One hypothesis for the role of ethylene in the promotion of herbicide-induced cell death is based on a potentially toxic by-product of ethylene biosynthesis. During ACC oxidase action, cyanide is formed in stoichiometric amounts with ethylene and CO_2 (Abeles et al. 1992). Normally, plant tissues have ample capacity to detoxify HCN to β-cyanoalanine by β-cyanoalanine synthase (β-CAS). Cyanide is a known inhibitor of metalloenzymes such as cytochrome c oxidase, Rubisco or catalase. It has been suggested (Grossmann and Kwiatkowski 1993; Grossmann 1996) that HCN may play a role in certain stress situations leading to highly elevated ethylene production, e.g., herbicide phytotoxicity and pathogen-induced hypersensitive response. It is, however, evident that cyanide production is not a general explanation for ethylene-mediated cell death. Mutant studies revealed that (high) ethylene production *per se* is not stimulatory for cell death, but under specific circumstances where ethylene synthesis is highly elevated and ethylene sensitivity is compromised, detoxification of cyanide by β-CAS may become compromised since the β-CAS gene is strongly ethylene-inducible (J. Vahala, J. Kangasjärvi unpubl. results). If plant ethylene sensitivity is decreased concurrently with highly enhanced ethylene synthesis, induced for example, by ozone, the deficient induction of β-CAS may compromise the efficient HCN detoxification leading to necrotic cell death. Experimental evidence indicating such has been obtained with transgenic birch trees that have been made ethylene insensitive (J. Vahala, J. Kangasjärvi, unpubl. results).

4.6 Roles of Signals of the Jasmonate Family under Wounding as Well as Herbivore and Pathogen Attack

Genetic evidence for the role of jasmonate in plant defense has been obtained from *Arabidopsis* mutants affected in jasmonate biosynthesis or signalling. The jasmonate-insensitive mutant *coi1* shows reduced growth of a virulent strain of *P. syringae* pv. *tomato* in the leaves, but exhibits enhanced susceptibility to necrotrophic fungi such as *Alternaria brassicicola* and *Botrytis cinerea* (Berger 2002). In addition, *jar1* and the triple mutant *fad3fad7fad8* (deficient in linolenic acid, and thus also jasmonate biosynthesis) exhibit increased susceptibility towards usually non-pathogenic soil microbes (*Pythium* ssp; Berger 2002). Finally, certain oxylipins, linoleic acid 9- and 13-ketodienes, which accumulated in *Pseudomonas*-infected *Arabidopsis* leaves, were able to induce expression of defence genes (e.g., *GST1*), but also contributed to host cell death (Vollenweider et al. 2000).

Another type of interaction of plants with other organisms, insect herbivores, also involves jasmonates in a very central role (for review, Kessler and Baldwin 2002). Activation of defense genes in herbivory is a result of the action of various signal molecules that are formed as the result of herbivore attack or other kind of mechanical wounding. These signals act both locally and systematically. Further studies have demonstrated that in herbivory, in a similar way as in pathogen defense, several different interactions of plant signalling routes are involved in a very specific way depending on the attacking species. For example, feeding of *Nicotiana attenuata* leaves by *Manduca sexta* larvae, which are adapted to feeding in *Nicotiana*, upregulated ethylene biosynthesis, which in turn prevented upregulation of the JA-dependent nicotine biosynthesis (Kahl et al. 2000). Furthermore, it was shown that the herbivory-induced ethylene reduced the JA-dependent nicotine accumulation by affecting the transcript levels of the gene encoding putrescine *N*-methyl transferase, a rate-limiting enzyme of nicotine biosynthesis (Winz and Baldwin 2001). However, the ethylene induced by feeding of *N. attenuata* by *Manduca sexta* did not affect the genes of jasmonate biosynthesis (Ziegler et al. 2001). Similarly, it was shown that ethylene and salicylate decrease and jasmonate increases defense responses of *Arabidopsis* to the generalist herbivore Egyptian cotton worm (*Spodopter littoralis*); ethylene and SA signalling mutants were more resistant to feeding by the larvae and jasmonate mutants more sensitive (Stotz et al. 2000, 2002). The interactions are even more complicated, however. It was shown, similarly as above, that SA increased the susceptibility of *Arabidopsis* also to the larvae of cabbage looper (*Trichoplusia ni*). However, the systemic, SA-mediated HR caused by infection with *Pseudomonas syringae* pv. *maculicola* overrode the SA-mediated increase in *T. ni* susceptibility and resulted in enhanced tolerance towards feeding by the larvae (Cui et al. 2002).

When the plant responses to insect herbivory were analyzed at the gene expression level, it became obvious that responses to mechanical wounding by the feeding insect are modulated in an interaction/species-specific manner. It has been shown that, for example, fatty acids/amino acids conjugates in the oral secretions of *Manduca sexta* were both necessary and sufficient for both herbivore-specific gene expression as well as the jasmonate burst. (Halitschke et al. 2001). Similarly, wounding of *Arabidopsis* leaves and feeding by the larvae of the cabbage butterfly (*Pieris rapae*) induced differential gene expression profiles when the plant response was analyzed with microarray hybridizations (Reymond et al. 2000; Stintzi et al. 2001).

4.7 Interactions Between Ethylene and Jasmonate Pathways

4.7.1 Activation of Stress-Responsive Genes

Ethylene and jasmonate modulate plant responses to a wide array of stress factors. Even though their signalling cascades can be considered as separate and linear pathways, they do not operate independently of each other, but rather interact in a complicated manner to regulate plant defense responses (Feys and Parker 2000; Thomma et al. 2001). Ethylene and jasmonate induce expression of several defense-related genes in a synergistic fashion (Penninckx et al. 1998; Norman-Setterblad et al. 2000). For example, the pathogenesis-related proteins PR1b and 5 are synergistically induced by MeJA and ethylene or by MeJA and salicylic acid in tobacco (Xu et al. 1994). In tomato, ethylene cooperatively stimulates the expression of wound and jasmonate-inducible proteinase inhibitor genes. Furthermore, it has been suggested that ethylene and jasmonate stimulate each other's biosynthesis in wound response (O'Donnell et al. 1996; Laudert and Weiler 1998). These potentiating effects may help to fine-tune defense gene expression against specific pathogens. The important outcome is that not all invading pathogens stimulate the same genes. However, ethylene can also inhibit jasmonate-mediated gene expression in certain systems (Rojo et al. 1999; Shoji et al. 2000; Ellis and Turner 2001; Winz and Baldwin 2001). It is possible that these contradictory observations reflect the fact that the interactions between the hormonal pathways depend on the nature of the external challenge.

Specific signature sets are apparent in the interactions of every signalling pathway. For example, while part of the ethylene-mediated signalling is antagonized by jasmonate during oxidative stress, a large portion of ethylene-regulated genes is synergistically induced by jasmonate (Penninckx et al. 1998; Norman-Setterblad et al. 2000; Schenk et al. 2000; Tuominen et al. 2002). The first group of genes is therefore thought to participate in the regulation of oxidative cell death, and the latter group to defense responses other than cell death. Similar kinds of dual signature sets between the ethylene and jasmonate pathways have been found during necrotrophic pathogen attack (Berger 2002). These pathogens induce in plants a burst of ethylene, which promotes cell death and is therefore beneficial for the pathogen (Thomma et al. 2001).

4.7.2 Complementing Roles in Rhizobacteria-Mediated Resistance

Defined strains of rhizobacteria trigger systemic pathogen resistance in *A. thaliana* as shown by Van Loon's group (see Chap 7, this Vol.). This type of resistance was termed "induced systemic resistance" (ISR) in order to differentiate it from the pathogen-induced and salicylate-mediated SAR (Van Loon et al. 1998; Shirasu and Schulze-Lefert 2000). During ISR, colonization with rhizobacteria, e.g., *Pseudomonas fluorescens* strain WCS417r, leads to an enhanced expression ("priming") of jasmonate-inducible genes. With the help of various biosynthesis and perception mutants for plant signal molecules, it was possible to demonstrate the absolute requirement of ISR for functional jasmonate and ethylene signalling even though the levels of the growth regulators were not influenced (Van Loon et al. 1998; Pieterse et al. 2001). It is highly interesting to note that SAR and ISR, each being effective against a different subset of pathogens, show additive effects against subsequent pathogen infections. Combining SAR and ISR in crop plants apparently is a promising way to increase plant health.

4.8 Roles of Ethylene and Jasmonate in Programmed Cell Death

4.8.1 The "Oxidative Cell Death Cycle"

Lesion formation in the HR, as well as in the related oxidative cell death, involves three separate processes: lesion initiation, lesion propagation and lesion containment. These processes are all under genetic control since lesion mimic or related mutants have been isolated in all three processes. Direct and integral involvement of ROS in the process has been demonstrated in several systems (Dangl et al. 1996; Lamb and Dixon 1997; Van Camp et al. 1998; Beers and McDowell 2001). The size of the lesion is determined by the balance between lesion propagation and containment. Recent results have suggested an integral involvement of hormonal signalling in determining the balance between propagation and containment; ethylene promotes lesion propagation and jasmonate counteracts this and promotes lesion containment.

The so-called oxidative cell death cycle was originally proposed by Van Camp et al. (1998) and was later modified by Overmyer et al. (2000) to specify the roles of ethylene and jasmonate. Ethylene promotes cell death in multiple pcd programs triggered by ozone, pathogens, in various mutations and in defined developmental stages. Increased ethylene synthesis or sensitivity triggers lesion propagation, and down-regulation of either ethylene synthesis or sensitivity promotes lesion containment (Fig. 4.5). Three ozone-sensitive *Arabidopsis* mutants, *rcd1*, *eto1* and *jar1*, which all show prolonged propagation and delayed containment of lesions, have a different molecular basis for their lesion formation. Both *rcd1* and *eto1* overproduce ethylene, however, through different regulatory systems, when compared to the Col-0 parent accession, which contributes to the lesion propagation. The jasmonate insensitive *jar1*,

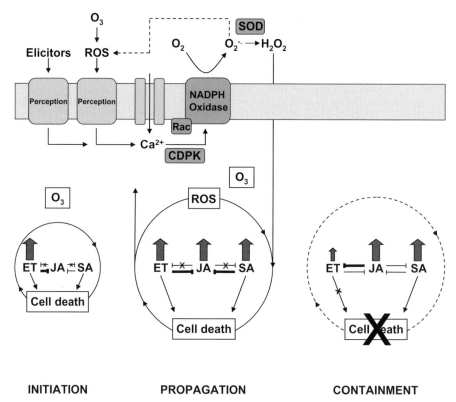

Fig. 4.5. Model of the initiation, propagation and containment phases of the oxidative cell death cycle in ozone-exposed and pathogen-treated plants (modified from Van Camp et al. 1998; Langebartels et al. 2002b; Tuominen et al. 2002). The propagation phase includes amplification of ROS accumulation by NADPH oxidase or other ROS-producing enzymes. *CDPK* Ca^{2+}-dependent protein kinase; *ET* ethylene; *JA* jasmonate; *Rac* low molecular weight GTPase; *ROS* reactive oxygen species; *SA* salicylic acid; *SOD* superoxide dismutase

however, has much lower basal ethylene evolution when compared to Col-0 and just a modest increase in ethylene synthesis as a response to ozone. However, lesion propagation is also ethylene dependent in *jar1*, and it has been shown to result from the hypersensitivity of *jar1* to ethylene (Tuominen et al 2002).

The mode of action for jasmonates in lesion containment seems to be through the regulation of ethylene sensitivity rather than synthesis. Exogenous jasmonate application to the *rcd1* mutant during the lesion propagation, when ethylene synthesis was still high, did not reduce, but, curiously, increased ethylene synthesis, and nevertheless resulted in lesion containment (Overmyer et al. 2000). It has recently been shown (Tuominen et al. 2002) that the jasmonate pathway modulates plant ethylene sensitivity. Thus, the inability of JAR1 to affect ethylene signalling in *jar1* mutant relieves ethylene signalling from the jasmonate regulation and results in ethylene hypersensitivity. Similarly, external jasmonate application in *rcd1* reduces its ethylene insensitivity, which leads into increased synthesis because of reduced negative feedback regulation, but the ethylene synthesized is not perceived. It was shown (Tuominen et al. 2002) that the jasmonate modulation of ethylene signalling occurs upstream of CTR1, but downstream of ethylene biosynthesis.

Several molecular mechanisms can be suggested for the antagonistic interaction between jasmonate and ethylene signalling, by which mechanisms jasmonate can affect and modulate plant ethylene sensitivity. It is known that ethylene sensitivity can be altered via transcriptional regulation of receptor levels (Ciardi et al. 2001). Intriguingly, the published microarray data show that at least one *Arabidopsis* ethylene-receptor (*ERS2*) is indeed jasmonate-inducible (Schenk et al. 2000), which could explain, at least partly, the effect of jasmonate on ethylene signalling; increased jasmonate-dependent synthesis of ERS2 receptors could down-regulate ethylene responses in a similar way as discussed above in ethylene signalling. Signal pathways for defense and cell death may also share common regulatory factors and compete for binding to the same target promoters. For example, a common *cis*-element has been found in the promoter of an EREBP/AP2-family transcription factor that relays both ethylene and JA signalling (Menke et al. 1999). However, which of the possible molecular mechanisms is involved in this interaction requires further experimental elucidation.

4.8.2 Cross-Talk with Other Signalling Molecules Coordinating Cell Death

The antagonistic roles and functions of jasmonate and ethylene in programmed cell death cannot, however, be described without the involvement of salicylic acid and ROS/NO, since these seem to be integrally involved in the regulation of cell death (Van Camp et al. 1998; Langebartels et al. 2002a,b).

Salicylic acid is a major phenolic compound involved in defense gene activation and initiation of cell death during oxidative stress (Alvarez 2000; Klessig et al. 2000; see also Chap. 5 by Métraux and Durner). It has been implicated in both local and systemic disease resistance towards pathogens. Salicylic acid functions both upstream and downstream of ROS, potentiates ROS toxicity and thereby contributes to cell death (Beers and McDowell 2001). Amplifying effects of salicylic acid for early defense responses such as the oxidative burst became apparent with the discovery of a rapid, transient peak of free salicylic acid in addition to the well-known late and sustained production (Draper 1997). This biphasic induction of salicylic acid has also been observed in response to elevated intracellular H_2O_2 levels in catalase antisense plants, indicating that H_2O_2 stimulates the production of its agonist, salicylic acid (Chamnongpol et al. 1998). Levels of free salicylic acid were induced before sustained ROS accumulation and cell death took place in ozone-sensitive *A. thaliana* and tobacco (Yalpani et al. 1994; Sharma and Davis 1997). On the other hand, salicylic acid levels as well as methyl salicylate emission were only weakly induced in ozone-tolerant tobacco (reviewed by Langebartels et al. 2002a).

Reactive oxygen species (ROS) play a crucial role in cell death processes in animals and plants. Pathogen infection, but also abiotic stress, activate (an) enzymatic complex(es) in plants responsible for a rapid and transient "oxidative burst" of apoplastic ROS accumulation (Lamb and Dixon 1997; Scheel 2002). Four enzymatic sources have been proposed as the basis for this burst in monocot and dicot species (reviewed in Bolwell 1999; Langebartels et al. 2002b; Scheel 2002): diamine and polyamine oxidases, oxalate oxidase, cell wall peroxidases and NADPH oxidase (Nox). A survey of the *Arabidopsis* genome revealed strong genetic evidence that plants employ homologs of mammalian Nox (Lambeth 2002) for ROS production. At least nine Nox isoforms were identified, of which three (*AtrbohD* to *F*) are expressed in leaf and root tissues (Keller et al. 1998; Torres et al. 1998, 2002). Functional Nox activity was found in microsomal and plasma membrane preparations of plants, and it occurred without assembly of an enzyme complex with cytosolic proteins as in the 'prototype' mammalian Nox2 (gp91phox) from macrophages. In addition, pathogen infection led to an increase in Nox expression and activity (Sagi and Fluhr 2001; Langebartels et al. 2002b; Simon-Plas et al. 2002), and harpin induced the expression of an *Arabidopsis* Nox homolog (Desikan et al. 1998), which points to an involvement of this enzyme in the pathogen response. In addition, it was implicated on the basis of inhibitor studies with the flavin oxidase inhibitor, diphenylene iodonium (DPI), that Nox is involved in the generation of the ozone-triggered oxidative burst in tobacco, *A. thaliana* and birch (Schraudner et al. 1998; Pellinen et al. 1999; Rao and Davis 1999; Overmyer et al. 2000). Two Nox genes, *NtrbohD* and *F*, were recently isolated from the ozone biomonitor plant, tobacco Bel W3. Their deduced amino acid sequences revealed close homology to the tomato and potato isoforms, as well as *Arabidopsis* Atrboh D and F (Langebartels et al. 2002b; Simon-Plas et al. 2002). While isoform F was constitutively present at basal levels in all tissues, *NtrbohD* transcripts as well as Nox activity were induced by ozone when plants were exposed in climate chambers and in the field.

Endogenous NO production in plants is activated under abiotic and biotic stress and interacts with ROS to trigger defense gene expression and cell death (Delledonne et al. 1998; Durner et al. 1998; Beligni and Lamattina 1999, 2001). NO emission responded extremely rapidly, within minutes of elicitation, in cryptogein-elicited epidermal cells of tobacco (Foissner et al. 2000). Recent results suggest that balanced NO and H_2O_2 levels trigger HR cell death, while spontaneous reaction of NO with O_2^- (at low SOD activity) rather leads to scavenging of NO to peroxynitrite (ONOO–), which is apparently less toxic in plants than in animals (Delledonne et al. 2001). Further investigations have to explore whether NO and H_2O_2 occur in the same clusters of cells that later exhibit programmed cell death. Given the importance of NO as a signal for defense and death, its effects as a phytotoxic air pollutant also have to be re-analyzed. For example, pre-treatment of pea plants with NO increased their sensitivity to ozone (Mehlhorn and Wellburn 1987). This result was not well understood before and may be based on the interaction of NO with ozone-triggered ROS production.

4.8.3 A Model for the Control of Cell Death

As shown in Fig. 4.5, ethylene, jasmonate and salicylic acid, and most probably also other signals, play a role as regulators in the production of ROS in pathogen-infected and ozone-exposed plants. Elicitor treatment or ozone exposure leads to a rapid burst (phase I) of ROS accumulation with minutes (cell cultures) or 1 to 2 h (intact plants), possibly by the activation of Nox activity. ROS accumulation can be visualized by diaminobenzidine staining for H_2O_2 and nitroblue tetrazolium staining for superoxide anions (Wohlgemuth et al. 2002). Re-current exposure to ozone preferentially induced ROS accumulation around the existing lesions, suggesting that activating signals are present in the border cells between dead and healthy tissue. A burst of ethylene biosynthesis (ACC content, ethylene emission) is stimulated in parallel in certain plant-pathogen interactions (Wang et al. 2002), catalase-deficient tobacco (Chamnongpol et al. 1998) and ozone-sensitive plants (Langebartels et al. 1991; Overmyer et al. 2000). It usually precedes the accumulation of other signals such as jasmonate and salicylate. A second, persistent burst of ROS accumulation (Phase II) is only found when ethylene is present above a certain (low) threshold level. Ethylene may either act by activating Nox activity or by blocking enzymes or low molecular weight compounds inhibiting Nox activity (Fig. 4.5). Activation could occur by phosphorylation in the stimulation of Nox activity as known for mammalian cells (Lambeth 2002). On the other hand, inactivation of Nox activity could be mediated by ethylene inhibition of the accumulation of polyamines, which are known Nox inhibitors in animal tissues (Walters et al. 2002). It was described before that ethylene and polyamines have opposite effects in ozone sensitivity of plants (Langebartels et al. 1991; Wellburn and Wellburn 1996) and mutually inhibit each other's biosynthesis (Abeles et al. 1992). In line with the antagonistic roles of ethylene and jasmonate, the latter was recently shown to promote polyamine accumulation in barley (Walters et al. 2002). Thus, the challenge for the next years will be to analyze the regulation of Nox and other potential ROS forming enzymes, e.g., cell wall peroxidases, by ethylene, jasmonate and other signals.

The model (modified and combined from Langebartels et al. 2002b and Tuominen et al. 2002) of Fig. 4.5 further predicts that plants can resist oxidative stress as long as the ethylene and salicylic acid pathways are not activated in excess, or when they are sufficiently inhibited by jasmonate. In addition, both ethylene and salicylate pathways have to be active for cell death to propagate. In the absence of the oxidative stress, the balance between the signalling pathways aims to inhibit activation of the cell death promoting effect of ethylene and salicylate. It seems that suppressing parts of the salicylate pathway by ethylene in the non-oxidative conditions could protect plants from spontaneous cell death under optimal growth conditions. During oxidative stress, the balance between the signalling pathways is altered towards ethylene and salicylate to allow cell death to occur. It seems that the biosynthetic kinetics of both ethylene and SA are critical for the regulation of the signalling balance and the concomitant cell death.

4.9 Ecotoxicological Implications of Ethylene and Jasmonate Release from Plants

Ethylene as well as jasmonate metabolites are released from leaves and roots of plants that are wounded or are under herbivore or pathogen attack. These metabolites can affect other organisms in the vicinity of the shoot or in the rhizosphere. For instance, the ethylene precursor ACC exuded from the roots of "stressed" plants may be taken up by bacteria in the rhizosphere. As detailed in Chapter 7 by Van Loon and Glick, various bacteria possess ACC deaminase activity, which cleaves the molecule to ammonia and a-ketobutyrate. By acting as a sink for ACC, these ACC deaminase-containing bacteria can control the level of ethylene in a plant, delay flower senescence, protect from flooding and heavy metal stress as well as ameliorate the damage from pathogen infestations (Glick et al. 1998).

Release of signal molecules like ethylene, methyl jasmonate, (Z)-jasmone and methyl salicylate from plants may affect neighbor plants as well as communication with other organisms (Farmer 2001). A bouquet of volatiles is typically released from ozone- and pathogen-treated plants; in the case of tobacco and tomato, ethylene is less than 1 % of total volatiles, while sesquiterpenes (25 %), C_6 aldehydes and alcohols (50 %) and methyl salicylate (24 %) make up the majority of coumpounds (Heiden et al. 1999). C_6 alcohols and aldehydes derive from the same precursor, the 13-hydroperoxide of linolenic acid, as jasmonate, but are formed by cleavage of 13-LOOH by hydroperoxide lyase (HPL) activity into one C_6 and one C_{12} (traumatic acid) molecule.

How can these emissions influence plants at the population level or other organisms in an ecosystem? First, growth may be controlled by sensors for "touch stress" and may be restricted when plants are constantly touching each other in dense populations. It was shown that the *Arabidopsis* ACC synthase isoform *At-ACS6* is rapidly and transiently induced by touch stress. Perception of other plants is affected in ETR1-mutated tobacco, which shows a so-called "crowding effect," non-controlled growth in the presence of neighbor plants (Knoester et al. 1998). Second, studies from our lab show that C_6 aldehydes and alcohols potentially increase the rate of cell death following ozone exposure or pathogen infection as they are toxic to plant cells (C. Langebartels, unpubl. results; see also Lam et al. 2001). In addition, the abundant C_6 aldehyde Z-3-hexenal induced phytoalexins in cotton and stress gene expression in *A. thaliana* (Zeringue 1992; Bate and Rothstein 1998), thus potentially raising resistance in plants. Third, airborne signalling involving methyl jasmonate or Z-jasmone may relate to defense against wounding organisms like herbivorous insects. Tomato plants cultivated together with sagebrush branches showed a strong activation of proteinase inhibitors as a result of the release of methyl jasmonate from sagebrush (Farmer and Ryan 1990). Recent experiments with *N. attenuata* and sagebrush at field sites showed that removal of sagebrush leaves led to the emission of 3R,7S-methyl jasmonate and a significant reduction of natural herbivore damage of *N. attenuata* in the vicinity of the wounded sagebrush (Karban et al. 2000). Both signal molecules, ethylene and jasmonate, are also able to trigger the formation of volatiles in various species, thus leading to amplifications of effects. For example, ethylene regulates the emission of aliphatic ester aroma volatiles in melons as shown in ACC oxidase antisense plants (Flores et al. 2002). The herbivore-induced ethylene burst suppressed nicotine accumulation, but not the release of volatiles from *N. attenuata* (Kahl et al. 2000). Jasmonate and related compounds induced volatile production in lima bean and *N. attenuata* (Koch et al. 1999; Halitschke et al. 2000). Some of these volatiles emitted from foliage may be protective against herbivore attack. Trees of the Brazil nut family, which emit increased levels of *S*-methyl methionine, are less colonized by wood boring beetles than low emitting trees (Berkov et al. 2000). Moreover, the C_6 ester hexenylacetate and other aliphatic esters of hexenol, being emitted from damaged tobacco leaves, deter female moths from laying eggs on these plants (De Moraes et al. 2001). Finally, jasmonate-dependent plant responses may include defense against herbivory at three trophic levels. Some of the jasmonate-inducible volatiles direct host-seeking parasitic insects, natural enemies of the herbivores, towards the plants being eaten by the caterpillars (Thaler 1999).

Several studies provide evidence that plant volatiles are able to influence the defense status of other plants in close vicinity. Methyl salicylate produced by a tobacco mosaic virus-infected tobacco plant activates SAR in a second plant kept in a different cuvette (Shulaev et al. 1997). *Vicia faba* plants exposed to (Z)-jasmone, a volatile jasmonate metabolite, showed elevated levels of monoterpenes and altered gene expression levels (Birkett et al. 2000). These results provide evidence that plant-derived signals are able to alter the disposition of plants towards biotic attack even when diluted in ambient air.

4.10 Concluding Remarks

For plants as sessile organisms, release of signal and defense compounds into the surrounding air and soil provides a way to communicate with organisms present in their vicinity. It seems that molecules like ethylene and jasmonate are primarily used to induce local defense responses and to regulate cell death at the attacked site. In a second episode, they may warn neighboring cells in a local resistance response, inform other organs and activate resistance systemically (auto-signalling), and finally communicate danger to neighbor plants and organisms (intra- and interspecific signalling; Farmer 2001). Depending on the concentration of these signals, which gradually decreases from local to systemic and air levels, ethylene and jasmonate play a role in amplification reactions leading to toxic effects, interfere with the growth of neighboring plants or induce defense against abiotic stress, pathogens and herbivores. An analogous communication system may exist in the soil compartment, with ethylene/ACC and jasmonate accumulating in roots, being effective in wounding and pathogen attack and, in addition, being released into the "communication zone," the rhizosphere. Using this system, soil organisms, but also other plants (via the 'mycorrhizal network') can perceive information on the actual state of a plant under abiotic or biotic stress. This multitude of defense reactions of plants towards abiotic and biotic stressors may generate metabolic trade-offs ultimately resulting in fitness costs (Baldwin et al. 2001; Heil 2001). Taken together, the above results from "mechanism-based" approaches lay the ground for ecotoxicologic studies. Future research will reveal whether environmental stress influences on regulatory cascades such as ethylene and jasmonate signal pathways will, through sequential propagation or interaction with other factors, lead to positive or negative consequences for plant populations in ecosystems.

Acknowledgements. The authors thank Susanne Berger and Ted Farmer for valuable discussions and critical reading of the manuscript. Work of our labs was supported by the Deutsche Forschungsgemeinschaft (SFB 607), Bayerisches Staatsministerium für Landesentwicklung und Umweltfragen, Bundesministerium für Forschung und Technologie, Fonds der Chemischen Industrie (to C.L.), Academy of Finland, the Finnish Centre of Excellence Program (to J.K.) and the EU-FAIR project TOMSTRESS (both labs).

References

Abeles FB, Morgan PW, Saltveit ME (eds) (1992) Ethylene in plant biology. Academic Press, New York

Alvarez ME (2000) Salicylic acid in the machinery of hypersensitive cell death and disease resistance. Plant Mol Biol 44:429–442

Avni A, Bailey BA, Mattoo AK, Anderson JD (1994) Induction of ethylene biosynthesis in *Nicotiana tabacum* by a *Trichoderma viride* xylanase is correlated to the accumulation of 1-aminocyclopropane-1-carboxylic acid (ACC) synthase and ACC oxidase transcripts. Plant Physiol 106:1049–1055

Baldwin IT, Halitschke R, Kessler A, Schittko U (2001) Merging molecular and ecological approaches in plant-insect interactions. Curr Opin Plant Biol 4:351–358

Barry CS, Llop-Tous MI, Grierson D (2000) The regulation of 1-aminocyclopropane-1-carboxylic acid synthase gene expression during the transition from system-1 to system-2 ethylene synthesis in tomato. Plant Physiol 123:979–986

Barry CS, Blume B, Bouzayen M, Cooper W, Hamilton AJ, Grierson D (1996) Differential expression of the 1-aminocyclopropane-1-carboxylate oxidase gene family of tomato. Plant J 9:525–535

Bate NJ, Rothstein SJ (1998) C_6-volatiles derived from the lipoxygenase pathway induce a subset of defense-related genes. Plant J 16:561–569

Beers EP, McDowell JM (2001) Regulation and execution of programmed cell death in response to pathogens, stress and developmental cues. Curr Opin Plant Biol 4:561–567

Beligni MV, Lamattina L (1999) Is nitric oxide toxic or protective? Trends Plant Sci 4:299

Beligni MV, Lamattina L (2001) Nitric oxide: a non-traditional regulator of plant growth. Trends Plant Sci 6:508–509

Bent AF, Innes RW, Ecker JR, Staskawicz BJ (1992) Disease development in ethylene-insensitive *Arabidopsis thaliana* infected with vurulent and avirulent *Pseudomonas* and *Xanthomonas* pathogens. Mol Plant Microbe Interact 5:372–378

Berger S (2002) Jasmonate-related mutants of *Arabidopsis* as tools for studying stress signaling. Planta 214:497–504

Berger S, Bell E, Mullet JE (1996) Two methyl jasmonate-insensitive mutants show altered expression of atvsp in response to methyl jasmonate and wounding. Plant Physiol 111:525–531

Berkov A, Meurer-Grimes B, Purzycki KL (2000) Do Lecythidaceae specialists (Coleoptera, Cerambycidae) shun fetid tree species? Biotropica 32:440–451

Birkett MA, Campbell CAM, Chamberlain K, Guerrieri E, Hick AJ, Martin JL, Matthes M, Napier JA, Pettersson J, Pickett J, Poppy GM, Pow EM, Pye BJ, Smart LE, Wadhams GH, Wadhams LJ, Woodcock CM (2000) New roles for cis-jasmone as an insect semiochemical and in plant defense. Proc Natl Acad Sci USA 97:9329–9334

Bleecker AB (1999) Ethylene perception and signaling: an evolutionary perspective. Trends Plant Sci 4:269–274

Bolwell GP (1999) Role of active oxygen species and NO in plant defence responses. Curr Opin Plant Biol 2:287–294

Brown KM (1997) Ethylene and abscission. Physiol Plant 100:567–576

Chamnongpol S, Willekens H, Moeder W, Langebartels C, Sandermann H, Van Montagu M, Inzé D, Van Camp W (1998) Defense activation and enhanced pathogen tolerance induced by H_2O_2 in transgenic tobacco. Proc Natl Acad Sci USA 95:5818–5823

Chang C, Stadler R (2001) Ethylene hormone receptor action in *Arabidopsis*. Bioessays 23:619–627

Chen YF, Randlett MD, Findell JL, Schaller GE (2002) Localization of the ethylene recep-
    tor ETR1 to the endoplasmic reticulum of *Arabidopsis*. J Biol Chem 277:19861–19866
Ciardi JA, Tieman DM, Lund ST, Jones JB, Stall RE, Klee HJ (2000) Response to *Xan-
    thomonas campestris* pv. *vesicatoria* in tomato involves regulation of ethylene recep-
    tor gene expression. Plant Physiol 123:81–92
Ciardi JA, Tieman DM, Jones JB, Klee HJ (2001) Reduced expression of the tomato ethyl-
    ene receptor gene *LeETR4* enhances the hypersensitive response to *Xanthomonas
    campestris* pv. *vesicatoria*. Mol Plant Microbe Interact 14:487–495
Clark KL, Larsen PB, Wang X, Chang C (1998) Association of the *Arabidopsis* CTR1 Raf-
    like kinase with the ETR1 and ERS ethylene receptors. Proc Natl Acad Sci USA
    95:5401–5406
Cohn J, Sessa G, Martin GB (2001) Innate immunity in plants. Curr Opin Immunol
    13:55–62
Cui J, Jander G, Racki LR, Kim PD, Pierce NE, Ausubel FM (2002) Signals involved in *Ara-
    bidopsis* resistance to *Trichoplusia ni* catepillars induced by viulent and avirulent
    strains of the phytopathogen *Pseudomonas syringae*. Plant Physiol 129:551–564
D'Agostino IB, Kieber JJ (1999) Phosphorelay signal transduction: the emerging family
    of plant response regulators. Trends Biochem Sci 24:452–456
Dangl JL, Jones JDG (2001) Plant pathogens and integrated defence responses to infec-
    tion. Nature 411:826–833
Dangl JL, Dietrich RA, Richberg MH (1996) Death don't have no mercy: cell death pro-
    grams in plant-microbe interactions. Plant Cell 8:1793–1807
Del Pozo O, Lam E (1998) Caspases and programmed cell death in the hypersensitive
    response of plants to pathogens. Curr Biol 8:1129–1132
Delledonne M, Xia Y, Dixon RA, Lamb C (1998) Nitric oxide functions as a signal in plant
    disease resistance. Nature 394:585–588
Delledonne M, Zeier J, Marocco A, Lamb C (2001) Signal interactions between nitric
    oxide and reactive oxygen intermediates in the plant hypersensitive disease resis-
    tance response. Proc Natl Acad Sci USA 98:13454–13459
De Moraes CM, Mescher MC, Tumlinson JH (2001) Caterpillar-induced noctunal plant
    volatiles repel nonspecific females. Nature 410:577–580
Desikan R, Burnett EC, Hancock JT, Neill SJ (1998) Harpin and hydrogen peroxide
    induce the expression of a homologue of gp91-*phox* in *Arabidopsis thaliana* suspen-
    sion cultures. J Exp Bot 49:1767–1771
Dion M, Chamberland H, St Michel C, Plante M, Darveau A, Lafontaine JG, Brisson LF
    (1997) Detection of a homologue of bcl-2 in plant cells. Biochem Cell Biol 75:457–461
Draper J (1997) Salicylate, superoxide synthesis and cell suicide in plant defence. Trends
    Plant Sci 2:162–166
Durner J, Wendehenne D, Klessig DF (1998) Defense gene induction in tobacco by nitric
    oxide, cyclic GMP, and cyclic ADP-ribose. Proc Natl Acad Sci USA 95:10328–10333
Ellis C, Turner JG (2001) The *Arabidopsis* mutant cev1 has constitutively active jas-
    monate and ethylene signal pathways and enhanced resistance to pathogens. Plant
    Cell 13:1025–1033
Ellis C, Karafyllidis I, Wasternack C, Turner JG (2002) The *Arabidopsis* mutant *cev1* links
    cell wall signaling to jasmonate and ethylene responses. Plant Cell 14:1557–1566
Elstner EF, Osswald W, Youngman RJ (1985) Basic mechanisms of pigment bleaching and
    loss of structural resistance in spruce (*Picea abies*) needles: advances in phytomed-
    ical diagnostics. Experientia 41:591–597
Farmer EE (2001) Surface-to-air signals. Nature 411:854–856
Farmer EE, Ryan CA (1990) Interplant communication: airborne methyl jasmonate
    induces synthesis of proteinase inhibitors in plant leaves. Proc Natl Acad Sci USA
    87:7713–7716

Felix G, Grosskopf DG, Regenass M, Boller T (1991) Rapid changes of protein phospho-
   rylation are involved in transduction of the elicitor signal in plant cells. Proc Natl
   Acad Sci USA 88:8831–8834
Feys B, Parker JE (2000) Interplay of signaling pathways in plant disease resistance.
   Trends Genet 16:449–455
Flores F, Yahyaoui FE, de Billerbeck G, Romojaro F, Latché A, Bouzayen M, Pech JC,
   Ambid C (2002) Role of ethylene in the biosynthetic pathway of aliphatic ester aroma
   volatiles in Charentais cantaloupe melons. J Exp Bot 53:201–206
Foissner I, Wendehenne D, Langebartels C, Durner J (2000) In vivo imaging of an elici-
   tor-induced nitric oxide burst in tobacco. Plant J 23:817–824
Fukuda H (2000) Programmed cell death of tracheary elements as a paradigm in plants.
   Plant Mol Biol 44:245–253
Galston AW, Kaur-Sawhney R, Altabella T, Tiburcio AF (1997) Plant polyamines in repro-
   ductive activity and response to abiotic stress. Bot Acta 110:197–207
Glick BR, Penrose DM, Li J (1998) A model for the lowering of plant ethylene concentra-
   tions by plant growth-promoting bacteria. J Theor Biol 190:63–68
Goodman RN, Novacky AJ (1994) The hypersensitive reaction in plants to pathogens: a
   resistance phenomenon. American Phytopathological Society Press, St Paul
Grant JJ, Loake GJ (2000) Role of reactive oxygen intermediates and cognate redox signaling in disease resistance. Plant Physiol 124:21–29
Grossmann K (1996) A role for cyanide, derived from ethylene biosynthesis, in the development of stress symptoms. Physiol Plant 97:772–775
Grossmann K, Kwiatkowski J (1993) Selective induction of ethylene and cyanide biosynthesis appears to be involved in the selectivity of the herbicide quinclorac between
   rice and barneygrass. J Plant Physiol 142:457–466
Halitschke R, Kessler A, Kahl J, Lorenz A, Baldwin IT (2000) Ecophysiological comparison of direct and indirect defenses in Nicotiana attenuata. Oecologica 124:408–417
Halitschke R, Schittko U, Pohnert G, Boland W, Baldwin IT (2001) Molecular interactions
   between the specialist herbivore Manduca sexta (Lepidoptera, Sphingiae) and its natural host Nicotiana attenuata. III. Fatty acid-amino acid conjugates in herbivore oral
   secretions are neccessary and sufficient for herbivore-specific plant responses. Plant
   Physiol 125:711–717
Hamberg M, Gardner HW (1992) Oxylipin pathway to jasmonates: biochemistry and
   biological significance. Biochim Biophys Acta 1165:1–18
Heath MC (2000) Hypersensitive response-related death. Plant Mol Biol 44:321–334
Heiden AC, Hoffmann T, Kahl J, Kley D, Klockow D, Langebartels C, Mehlhorn H, San-
   dermann H, Schraudner M, Schuh G, Wildt J (1999) Emission of volatile organic com-
   pounds from ozone-exposed plants. Ecol Appl 9:1160–1167
Heil M (2001) Induced systemic resistance (ISR) against pathogens – a promising field
   for ecological research. Perspect Plant Ecol Evol Syst 4:65–79
Heil M, Baldwin IT (2002) Fitness costs of induced resistance: emerging experimental
   support for a slippery concept. Trends Plant Sci 7:61–67
Howe GA, Schilmiller AL (2002) Oxylipin metabolism in response to stress. Curr Opin
   Plant Biol 5:230–236
Howe GA, Lee GI, Itoh A, Li L, DeRocher AE (2000) Cytochrome P450-dependent metab-
   olism of oxylipins in tomato. Cloning and expression of allene oxide synthase and
   fatty acid hydroperoxide lyase. Plant Physiol 123:711–724
Hua J, Meyerowitz EM (1998) Ethylene responses are negatively regulated by a receptor
   gene family in Arabidopsis thaliana. Cell 94:261–271
Inzé D, Van Montagu M (1995) Oxidative stress in plants. Curr Opin Biotechnol
   6:153–158

John P (1997) Ethylene biosynthesis: the role of 1-aminocyclopropane-1-carboxylate (ACC) oxidase, and its possible evolutionary origin. Physiol Plant 100:583–592

Johnson PR, Ecker JR (1998) The ethylene gas signal transduction pathway: a molecular perspective. Annu Rev Genet 32:227–254

Kahl J, Siemens DH, Aerts RJ, Gäbler R, Kühnemann F, Preston CA, Baldwin IT (2000) Herbivore-induced ethylene suppresses a direct defense but not a putative indirect defense against an adapted herbivore. Planta 210:336–342

Kakimoto T (1998) Cytokinin signaling. Curr Opin Plant Biol 1:399–403

Kangasjärvi J, Tuominen H, Overmyer K (2001) Ozone-induced cell death. Reactive oxygen species as signal molecules regulating cell death. In: Huttunen S, (ed) Trends in European forest tree physiology research. Kluwer, Dordrecht, pp 157–166

Karban R, Baldwin IT (1997) Induced responses to herbivory. University of Chicago Press, Chicago

Karban R, Baldwin IT, Baxter KJ, Laue JG, Felton GW (2000) Communication between plants: induced resistance in wild tobacco plants following clipping of neighboring sagebrush. Oecologia 125:66–71

Keller T, Damude HG, Werner D, Doerner P, Dixon RA, Lamb C (1998) A plant homolog of the neutrophil NADPH oxidase gp91[phox] subunit gene encodes a plasma membrane protein with Ca^{2+} binding motifs. Plant Cell 10:255–266

Kende H (1993) Ethylene biosynthesis. Annu Rev Plant Physiol Plant Mol Biol 44:283–307

Kessler A, Baldwin IT (2002) Plant responses to insect herbivory: the emerging molecular analysis. Annu Rev Plant Biol 53:299–328

Kidd VJ, Lahti JM, Teitz T (2000) Proteolytic regulation of apoptosis. Semin Cell Dev Biol 11:191–201

Kim YS, Choi D, Lee MM, Lee SH, Kim WT (1998) Biotic and abiotic stress-related expression of 1-aminocyclopropane-1-carboxylate oxidase gene family in *Nicotiana glutinosa* L. Plant Cell Physiol 39:565–573

Klessig DF, Durner J, Noad R, Navarre DA, Wendehenne D, Kumar D, Zhou JM, Shah J, Zhang S, Kachroo P, Trifa Y, Pontier D, Lam E, Silva H (2000) Nitric oxide and salicylic acid signaling in plant defense. Proc Natl Acad Sci USA 97:8849–8855

Knoester M, Bol JF, Van Loon LC, Linthorst HJM (1995) Virus-induced gene expression for enzymes of ethylene biosynthesis in hypersensitively reacting tobacco. Mol Plant Microbe Interact 8:177–180

Knoester M, Van Loon LC, Van den Heuvel J, Hennig J, Bol JF, Linthorst HJM (1998) Ethylene-insensitive tobacco lacks nonhost resistance against soil-borne fungi. Proc Natl Acad Sci USA 95:1933–1937

Koch T, Krumm T, Jung V, Engelberth J, Boland W (1999) Differential induction of plant volatile biosynthesis in the lima bean by early and late intermediates of the octadecanoid-signaling pathway. Plant Physiol 121:153–162

Korthout HA, Berecki G, Bruin W, van Duijn B, Wang M (2000) The presence and subcellular localization of caspase 3-like proteinases in plant cells. FEBS Lett 475:139–144

Lam E, Kato N, Lawton M (2001) Programmed cell death, mitochondria and the plant hypersensitive response. Nature 411:848–853

Lamb C, Dixon RA (1997) The oxidative burst in plant disease resistance. Annu Rev Plant Physiol Plant Mol Biol 48:251–275

Lambeth JD (2002) Nox/Duox family of nicotinamide adenine dinucleotide (phosphate) oxidases. Curr Opin Hematol 9:11–17

Langebartels C, Kerner K, Leonardi S, Schraudner M, Trost M, Heller W, Sandermann H (1991) Biochemical plant responses to ozone. I. Differential induction of polyamine and ethylene biosynthesis in tobacco. Plant Physiol 95:882–889

Langebartels C, Schraudner M, Heller W, Ernst D, Sandermann H (2002a) Oxidative stress and defense reactions in plants exposed to air pollutants and UV-B radiation. In: Inzé D, Van Montagu M (eds) Oxidative stress in plants. Taylor and Francis, London, pp 105–135

Langebartels C, Wohlgemuth H, Kschieschan S, Grün S, Sandermann H (2002b) Oxidative burst and cell death in ozone-exposed plants. Plant Physiol Biochem 40:567–575

Lashbrook CC, Tieman DM, Klee HJ (1998) Differential regulation of the tomato *ETR* gene familiy throughout plant development. Plant J 15:243–252

Laudert D, Weiler EW (1998) Allene oxide synthase: a major control point in *Arabidopsis thaliana* octadecanoid signaling. Plant J 15:675–684

Lelièvre J-M, Latché A, Jones B, Bouzayen M, Pech J-C (1997) Ethylene and fruit ripening. Physiol Plant 101:727–739

Levine A, Pennell RI, Alvarez ME, Palmer R, Lamb C (1996) Calcium-mediated apoptosis in a plant hypersensitive disease resistance response. Curr Biol 6:427–437

Liechti R, Farmer EE (2002) The jasmonate pathway. Science 296:1649–1650

Lincoln JE, Campbell AD, Oetiker J, Rottmann WH, Oeller PW, Shen NF, Theologis A (1993) LE-ACS4, a fruit ripening and wound-induced 1-aminocyclopropane-1-carboxylate synthase gene of tomato (*Lycopersicon esculentum*). J Biol Chem 268:19422–19430

Liu D, Li N, Dube S, Kalinski A, Herman E, Mattoo AK (1993) Molecular characterization of a rapidly and transiently wound-induced soybean (*Glycine max* L.) gene encoding 1-aminocyclopropan-1-carboxylate synthase. Plant Cell Physiol 34:1151–1157

Llop-Tous I, Barry CS, Grierson D (2000) Regulation of ethylene biosynthesis in response to pollination in tomato flowers. Plant Physiol 123:971–978

Lohrmann J, Harter K (2002) Plant two-component signaling systems and the role of response regulators. Plant Physiol 128:363–369

McConkey D, Orrenius S (1994) Signal transduction pathways to apoptosis. Trends Cell Biol 4:370–375

Mehlhorn H, Wellburn AR (1987) Stress ethylene formation determines plant sensitivity to ozone. Nature 327:417–418

Menke FLH, Champion A, Kijne JW, Memelink J (1999) A novel jasmonate- and elicitor-responsive element in the periwinkle secondary metabolite biosynthetic gene *Str* interacts with a jasmonate- and elicitor-inducible AP2-domain transcription factor, ORCA2. EMBO J 18:4455–4463

Meskiene I, Hirt H (2000) MAP kinase pathways: molecular plug-and-play chips for the cell. Plant Mol Biol 42:791–806

Mittler R, Rizhsky L (2000) Transgene-induced lesion mimic. Plant Mol Biol 44:335–344

Moeder W, Barry CS, Tauriainen A, Betz C, Tuomainen J, Utriainen M, Grierson D, Sandermann H, Langebartels C, Kangasjärvi J (2002) Ethylene synthesis regulated by bi-phasic induction of 1-aminocyclopropane-1-carboxylic acid synthase and 1-aminocyclopropane-1-carboxylic acid oxidase genes is required for H_2O_2 accumulation and cell death in ozone-exposed tomato. Plant Physiol 130:1918–1926

Morel J-B, Dangl JL (1997) The hypersensitive response and the induction of cell death in plants. Cell Death Differ 4:671–683

Morgan PW, Drew MC (1997) Ethylene and plant responses to stress. Physiol Plant 100:620–630

Moriarty F (1988) Ecotoxicology. Hum Toxicol 7:437–441

Nakatsuka A, Murachi S, Okunishi H, Shiomi S, Nakano R, Kubo Y, Inaba A (1998) Differential expression and internal feedback regulation of 1-aminocyclopropane-1-carboxylate synthase, 1-aminocyclopropane-1-carboxylate oxidase, and ethylene receptor genes in tomato fruit during development and ripening. Plant Physiol 118:1295–1305

Norman-Setterblad C, Vidal S, Palva ET (2000) Interacting signal pathways control defense gene expression in *Arabidopsis* in response to cell wall-degrading enzymes from *Erwinia carotovora*. Mol Plant Microbe Interact 13:430–438

O'Donnell PJ, Calvert C, Atzorn R, Wasternack C, Leyser HMO, Bowles DJ (1996) Ethylene as a signal mediating the wound response of tomato plants. Science 274:1914–1917

Oetiker JH, Olson DC, Shiu OY, Yang SF (1997) Differential induction of seven 1-aminocyclopropane-1-carboxylate synthase genes by elicitor in suspension cultures of tomato (*Lycopersicon esculentum*). Plant Mol Biol 34:275–286

Overmyer K, Tuominen H, Kettunen C, Betz C, Langebartels C, Sandermann H, Kangasjärvi J (2000) The ozone-sensitive *Arabidopsis rcd1* mutant reveals opposite roles for ethylene and jasmonate signaling pathways in regulating superoxide-dependent cell death. Plant Cell 12:1849–1862

Overmyer K, Pellinen R, Kuittinen T, Saarma M, Kangasjärvi J (2003) Ozone induced programmed cell death in the *radical-induced cell death1* mutant of *Arabidopsis*. Plant J (submitted)

Parkinson JS, Kofoid EC (1992) Communication modules in bacterial signaling proteins. Annu Rev Genet 26:71–112

Pellinen R, Palva T, Kangasjärvi J (1999) Subcellular localization of ozone-induced hydrogen peroxide production in birch (*Betula pendula*) leaf cells. Plant J 20:349–356

Pennell RI, Lamb C (1997) Programmed cell death in plants. Plant Cell 9:1157–1168

Penninckx IAMA, Thomma BPHJ, Buchala A, Métraux JP, Broekaert WF (1998) Concomitant activation of jasmonate and ethylene response pathways is required for induction of a plant defensin gene in *Arabidopsis*. Plant Cell 10:2103–2113

Pierson LSI, Wood DW, Pierson EA (1998) Homoserine lactone-mediated gene regulation in plant-associated bacteria. Annu Rev Phytopathol 36:207–225

Pieterse CMJ, Van Loon LC (1999) Salicylic acid-independent plant defence pathways. Trends Plant Sci 4:52–58

Pieterse CMJ, Ton J, Van Loon LC (2001) Cross-talk between defence signalling pathways: boost or burden? AgBioTechNet 3:1–8

Rancé I, Fournier J, Esquerré-Tugayé MT (1998) The incompatible interaction between *Phytophthora parasitica* var. *nicotianae* race 0 and tobacco is suppressed in transgenic plants expressing antisense lipoxygenase sequences. Proc Natl Acad Sci USA 95:6554–6559

Rao MV, Davis KR (1999) Ozone-induced cell death occurs via two distinct mechanisms in *Arabidopsis*: the role of salicylic acid. Plant J 17:603–614

Rao MV, Davis KR (2001) The physiology of ozone-induced cell death. Planta 213:682–690

Rao MV, Koch JR, Davis KR (2000) Ozone: a tool for probing programmed cell death in plants. Plant Mol Biol 44:345–358

Reymond P, Farmer EE (1998) Jasmonate and salicylate as global signals for defense gene expression. Curr Opin Plant Biol 1:404–411

Reymond P, Weber H, Damond M, Farmer EE (2000) Differential gene expression in response to mechanical wounding and insect feeding in Arabidopsis. Plant Cell 12:707–719

Robertson JD, Orrenius S (2000) Molecular mechanisms of apoptosis induced by cytotoxic chemicals. Crit Rev Toxicol 30:609–627

Rojo E, León J, Sánchez-Serrano JJ (1999) Cross-talk between wound signalling pathways determines local versus systemic gene expression in *Arabidopsis thaliana*. Plant J 20:135–142

Rottmann WH, Peter GF, Oeller PW, Keller JA, Shen NF, Nagy BP, Taylor LP, Campbell AD, Theologis A (1991) 1-Aminocyclopropane-1-carboxylate synthase in tomato is encoded by a multigene family whose transcription is induced during fruit and floral senescence. J Mol Biol 222:937–961

Rustérucci C, Montillet JL, Agnel JP, Battesti C, Alonso B, Knoll A, Bessoule JJ, Etienne P, Suty L, Blein JP, Triantaphylidès C (1999) Involvement of lipoxygenase-dependent production of fatty acid hydroperoxides in the development of the hypersensitive cell death induced by cryptogein on tobacco leaves. J Biol Chem 274:36446–36455

Sagi M, Fluhr R (2001) Superoxide production by plant homologues of the gp91[phox] NADPH oxidase. Modulation of activity by calcium and by tobacco mosaic virus infection. Plant Physiol 126:1281–1290

Sakakibara H, Hayakawa A, Deji A, Gawronski SW, Sugiyama T (1999) His-Asp phospho-transfer possibly involved in the nitrogen signal transduction mediated by cytokinin in maize: Molecular cloning of cDNAs for two-component regulatory factors and demonstration of phosphotransfer activity in vitro. Plant Mol Biol 41:563–573

Sakakibara H, Taniguchi M, Sugiyama T (2000) His-Asp phosphorelay signaling: a com-munication avenue between plants and their environment. Plant Mol Biol 42:273–278

Sandermann H (1994) Higher plant metabolism of xenobiotics: the 'green liver' concept. Pharmacogenetics 4:225–241

Sandermann H, Ernst D, Heller W, Langebartels C (1998) Ozone: an abiotic elicitor of plant defense reactions. Trends Plant Sci 3:47–50

Scheel D (2002) Oxidative burst and the role of reactive oxygen species in plant-pathogen interactions. In: Inzé D, Van Montagu M (eds) Oxidative stress in plants. Taylor and Francis, London, pp 137–153

Schenk PM, Kazan K, Wilson I, Anderson JP, Richmond T, Somerville SC, Manners JM (2000) Coordinated plant defense responses in Arabidopsis revealed by microarray analysis. Proc Natl Acad Sci USA 97:11655–11660

Schraudner M, Moeder W, Wiese C, Van Camp W, Inzé D, Langebartels C, Sandermann H (1998) Ozone-induced oxidative burst in the ozone biomonitor plant, tobacco Bel W3. Plant J 16:235–245

Seiler JP (2002) Pharmacodynamic activity of drugs and ecotoxicology – can the two be connected? Toxicol Lett 131:105–115

Sharma YK, Davis KR (1997) The effects of ozone on antioxidant responses in plants. Free Radical Biol Med 23:480–488

Shirasu K, Schulze-Lefert P (2000) Regulators of cell death in disease resistance. Plant Mol Biol 44:371–385

Shiu OY, H OJ, Yip WK, Yang SF (1998) The promoter of *LE-ACS7*, an early flooding-induced 1-amiocyclopropane-1-carboxylate synthase gene of tomato, is tagged by a *Sol3* transposon. Proc Natl Acad Sci USA 95:10334–10339

Shoji T, Nakajima K, Hashimoto T (2000) Ethylene suppresses jasmonate-induced gene expression in nicotine biosynthesis. Plant Cell Physiol 41:1072–1076

Shulaev V, Silverman P, Raskin I (1997) Airborne signalling by methyl salicylate in plant pathogen interactions. Nature 385:718–721

Simon-Plas F, Elmayan T, Blein JP (2002) The plasma membrane oxidase NtrbohD is responsible for AOS production in elicited tobacco cells. Plant J 31:137–147

Solano R, Ecker JR (1998) Ethylene gas: perception, signaling and response. Curr Opin Plant Biol 1:393–398

Solano R, Stepanova A, Chao Q, Ecker JR (1998) Nuclear events in ethylene signaling: a transcriptional cascade mediated by ETHYLENE-INSENSITIVE3 and ETHYLENE-RESPONSE-FACTOR1. Genes Dev 2:3703–3714

Spanu P, Grosskopf DG, Felix G, Boller T (1994) The apparent turnover of 1-aminocyclo-propane-1-carboxylate synthase in tomato cells is regulated by protein phosphoryla-tion and dephosphorylation. Plant Physiol 106:529–535

Stakman EC (1915) Relation between *Puccinia graminis* and plants highly resistant to its attack. J Agric Res 4:193–201

Staswick PE, Tiryaki I, Rowe ML (2002) Jasmonate response locus *JAR1* and several related *Arabidopsis* genes encode enzymes of the firefly luciferase superfamily that show activity on jasmonic, salicylic, and indole-3-acetic acids in an assay for adenylation. Plant Cell 14:1–11

Stepanova AN, Ecker JR (2000) Ethylene signaling: from mutants to molecules. Curr Opin Plant Biol 3:353–360

Stintzi A, Weber H, Reymond P, Browse J, Farmer EE (2001) Plant defense in the absence of jasmonic acid: the role of cyclopentenones. Proc Natl Acad Sci USA 98:12837–12842

Stotz HU, Pittendrihh BR, Kroymann J, Weniger K, Fritsche AB, Mitchell-Olds T (2000) Induced plant defense responses against chewing insects. Ethylene signaling reduces resistance of *Arabidopsis* against Egytian cotton worm but not Diamondback moth. Plant Physiol 124:1007–1017

Stotz HU, Koch T, Biedermann A, Weniger K, Boland W, Mitchell-Olds T (2002) Evidence for regulation of resistance in *Arabidopsis* to Egyptian cotton worm by salicylic and jasmonic acid signaling pathways. Planta 214:648–652

Tatsuki M, Mori H (1999) Rapid and transient expression of 1-aminocyclopropane-1-carboxylate synthase isogenes by touch and wound stimuli in tomato. Plant Cell Physiol 40:709–715

Tatsuki M, Mori H (2001) Phosphorylation of tomato 1-aminocyclopropane-1-carboxylic acid synthase, LE-ACS2, at the C-terminal region. J Biol Chem 276:28051–28057

Thaler JS (1999) Jasmonate-inducible defences cause increased parasitism of herbivores. Nature 399:686–688

Thomma BPHJ, Penninckx IAMA, Broekaert WF, Cammue BPA (2001) The complexity of disease signaling in *Arabidopsis*. Curr Opin Immunol 13:63–68

Tieman DM, Klee HJ (1999) Differential expression of two novel members of the tomato ethylene-receptor family. Plant Physiol 120:165–172

Tingey DT, Standley C, Field RW (1976) Stress ethylene evolution: a measure of ozone effects on plants. Atmos Environ 10:969–974

Torres M, Onouchi H, Hamada S, Machida C, Hammond-Kosack K, Jones J (1998) Six *Arabidopsis thaliana* homologues of the human respiratory burst oxidase (*gp91phox*). Plant J 14:365–370

Torres MA, Dangl JL, Jones JDG (2002) *Arabidopsis* gp91phox homologues *AtrbohD* and *AtrbohF* are required for accumulation of reactive oxygen intermediates in the plant defense response. Proc Natl Acad Sci USA 99:517–522

Tuomainen J, Betz C, Kangasjärvi J, Ernst D, Yin ZH, Langebartels C, Sandermann H (1997) Ozone induction of ethylene emission in tomato plants: Regulation by differential accumulation of transcripts for the biosynthetic enzymes. Plant J 12:1151–1162

Tuominen H, Overmyer K, Keinänen M, Kangasjärvi J (2003) Mutually antagonistic interactions between ethylene, jasmonic acid and salicylic acid in the regulation of ozone-induced cell death in *Arabidopsis*. Plant Cell (submitted)

Turner JG, Ellis C, Devoto A (2002) The jasmonate signal pathway. Plant Cell 14:S153-S164

Uren AG, O'Rourke K, Aravind L, Pisabarro MT, Seshagiri S, Koonin EV, Dixit VM (2000) Identification of paracaspases and metacaspases: two ancient families of caspase-like proteins, one of which plays a key role in MALT lymphoma. Mol Cell 6:961–967

Van Camp W, Van Montagu M, Inzé D (1998) H_2O_2 and NO: redox signals in disease resistance. Trends Plant Sci 3:330–334

Van Loon LC, Bakker PAHM, Pieterse CMJ (1998) Systemic resistance induced by rhizosphere bacteria. Annu Rev Phytopathol 36:453–483

Vogel J, Woeste K, Theologis A, Kieber J (1998) Recessive and dominant mutations in the ethylene biosynthetic gene *ACS5* of *Arabidopsis* confer cytokinin insensitivity and ethylene overproduction, respectively. Proc Natl Acad Sci USA 95:4766–4771

Vollenweider S, Weber H, Stolz S, Chételat A, Farmer EE (2000) Fatty acid ketodienes and fatty acid ketotrienes: Michael addition acceptors that accumulate in wounded and diseased *Arabidopsis* leaves. Plant J 24:467–476

von Tiedemann A (1997) Evidence for a primary role of active oxygen species in induction of host cell death during infection of bean leaves with *Botrytis cinerea*. Physiol Mol Plant Pathol 50:151–166

Walters D, Cowley T, Mitchell A (2002) Methyl jasmonate alters polyamine metabolism and induces systemic protection against powdery mildew infection in barley seedlings. J Exp Bot 53:747–756

Wang KLC, Li H, Ecker JR (2002) Ethylene biosynthesis and signaling networks. Plant Cell 14:S131–S151

Weber H, Vick BA, Farmer EE (1997) Dinor-oxo-phytodienoic acid: a new hexadecanoid signal in the jasmonate family. Proc Natl Acad Sci USA 94:10473–10478

Wellburn FAM, Wellburn AR (1996) Variable patterns of antioxidant protection but similar ethene emission differences in several ozone-sensitive and ozone-tolerant plant selections. Plant Cell Environ 19:754–760

Wilkinson JQ, Lanahan MB, Yen HC, Giovannoni JJ, Klee HJ (1995) An ethylene-inducible component of signal transduction encoded by Never-ripe. Science 270:1807–1809

Winz RA, Baldwin IT (2001) Molecular interactions between the specialist herbivore *Manduca sexta* (Lepidoptera, Sphingidae) and its natural host *Nicotiana attenuata*. IV. Insect-induced ethylene reduces jasmonate-induced nicotine accumulation by regulating putrescine *N*-methyltransferase. Plant Physiol 125:2189–2209

Wohlgemuth H, Mittelstrass K, Kschieschan S, Bender J, Weigel HJ, Overmyer K, Kangasjärvi J, Sandermann H, Langebartels C (2002) Activation of an oxidative burst is a general feature of sensitive plants exposed to the air pollutant ozone. Plant Cell Environ 25:717–726

Xie D, Feys B, James S, Nieto-Rostro M, Turner J (1998) COI1: An *Arabidopsis* gene required for jasmonate-regulated defense and fertility. Science 280:1091–1094

Xu L, Liu F, Lechner E, Genschik P, Crosby WL, Ma H, Peng W, Huang D, Xie D (2002) The SCF^COI1 ubiquitin-ligase complexes are required for jasmonate response in *Arabidopsis*. Plant Cell 14:1919–1935

Xu Y, Chang P, Liu D, Narasimhan M, Raghothama K, Hasegawa P, Bressan R (1994) Plant defense genes are synergistically induced by ethylene and methyl jasmonate. Plant Cell 6:1077–1085

Yalpani N, Enyedi AJ, León J, Raskin I (1994) Ultraviolet light and ozone stimulate accumulation of salicylic acid, pathogenesis-related proteins and virus resistance in tobacco. Planta 193:372–376

Yang SF, Hoffman NE (1984) Ethylene biosynthesis and its regulation in higher plants. Annu Rev Plant Physiol 35:155–189

Zeringue HJ (1992) Effects of C_6-C_{10} alkenals and alkanals on eliciting a defence response in the developing cotton boll. Phytochemistry 31:2305–2308

Ziegler J, Keinänen M, Baldwin IT (2001) Herbivore-induced allene oxide synthase transcripts and jasmonic acid in *Nicotiana attenuata*. Phytochemistry 8:729–738

5 The Role of Salicylic Acid and Nitric Oxide in Programmed Cell Death and Induced Resistance

Jean-Pierre Métraux and Jörg Durner

5.1 Plant Defense Responses

The ability to respond to environmental signals and challenges is essential to the survival of all organisms. Animals possess a sophisticated and unmatched immune system, which allows them to develop systemic, long-lasting and highly specific resistance to invading pathogens. Plants react to pathogen attack by activating elaborate defense mechanisms (Dangl and Jones 2001). The defense response is activated not only at the sites of infection, which are manifested as necrotic lesions (hypersensitive response), but also in neighboring and even distal uninfected parts of the plant, leading to systemic acquired resistance. Plant resistance is associated with activated expression of a large number of defense-related genes, whose products may play important roles in the restriction of pathogen growth and spread. During the past several years, a mounting body of evidence has accumulated, which indicates that salicylic acid (SA) acts as an endogenous signal for plant defense responses (Métraux 2001). Recent research on redox signalling has focused on nitric oxide (NO), which is involved in both physiological and pathophysiological conditions (Wendehenne et al. 2001). SA and NO, together with reactive oxygen species and other mediators, are constituents of a signalling network regulating plant defense responses.

5.1.1 The Hypersensitive Response: A Form of Programmed Cell Death

In plant-pathogen interactions, a susceptible plant cannot restrict pathogen growth and/or spread; thus, the pathogen often causes severe damage or even death of the entire plant. In contrast, a resistant plant is capable of deploying a variety of defense responses to prevent pathogen colonization. A key difference between resistant and susceptible plants is the timely recognition of the invading pathogen and the rapid and effective activation of host defenses. The

Ecological Studies, Vol. 170
H. Sandermann (Ed.)
Molecular Ecotoxicology of Plants
© Springer-Verlag Berlin Heidelberg 2004

activated defense response is frequently manifested in part as the so-called hypersensitive response (HR), which is characterized by necrosis at the sites of infection (resembling animal programmed cell death), and restriction of pathogen growth and spread (Dangl 1998; Shirasu and Schulze-Lefert 2000; Beers and McDowell 2001; Dangl and Jones 2001). Thus, plants use apoptosis in response to infection by pathogens to protect the whole plant survival. It should be noted, however, that HR development is not restricted to pathogen interactions. Recently, it has been shown that ozone-induced lesion development in tobacco and *Arabidopsis* is based on molecular mechanisms known to occur after pathogen attack (Sandermann et al. 1998; Sandermann 2000).

The development of HR is accompanied by the activation of many plant protectant and defense genes. Their products include glutathione *S*-transferases (GST), peroxidases, cell wall proteins and enzymes involved in phytoalexin biosynthesis, such as phenylalanine ammonia lyase (PAL) and chalcone synthase (Hammond-Kosack and Jones 1996; Ryals et al. 1996). Here, however, we want to focus on initiation of HR and signals regulating HR, with an emphasis on redox events and reactive oxygen species.

The induction of HR as a reaction to biotic and abiotic stresses requires the perception of a stimulus by a receptor and the subsequent involvement of second messengers and effector proteins to trigger an appropriate response (Dangl and Jones 2001). The activation of the HR is initiated by host recognition of race-specific (e.g., avirulence gene products) or non-specific signals (e.g., microbial proteins, glycoproteins, small peptides and oligosaccharides, etc.) (Schaller and Ryan 1996; Dangl and Jones 2001). The "gene-for-gene" paradigm is reminiscent of animal immune responses; these genes apparently control receptor-ligand interactions that activate a complex host response. Apparently, there are hundreds of resistance genes that mediate the recognition of specific fungal, bacterial, viral or nematode pathogen strains (Dangl and Jones 2001). If either the host or the pathogen lacks the corresponding *avr* or *R* gene, then the plant-pathogen interaction results in disease. Pathogen-derived *avr* gene products are delivered to intercellular spaces or directly inside plant cells, where they interact with the products of plant *R* genes (Yu et al. 2000).

The R proteins are either transmembrane or intracellular proteins that initiate HR and subsequent downstream signal-transduction cascades upon ligand binding (Nürnberger et al. 1994; Whitham et al. 1994; Wendehenne et al. 1995). *R* gene products share structural motifs, indicating that similar pathways might control resistance to diverse pathogens: nucleotide-binding sites (NBS), leucine-zipper motives and leucine-rich repeat (LRR) domains, functional domains of mammalian IL-1 receptor (IL-1R) and the *Drosophila* Toll proteins (TIR [Toll/IL-1R]) domain, ankyrins and many others (Cohn et al. 2001; Dangl and Jones 2001). The presence of an NBS region suggests that these proteins may play a role in the activation of a kinase or as a G protein. The leucine-zipper region, usually present in R proteins of this class, is likely to be involved in protein-protein interactions. The LRR region present in many *R* genes is a major factor responsible for the specificity of pathogen recognition (Ellis et al. 1999). However, as mentioned above, not all *R* genes possess LRR regions, and it is, therefore, very likely that there are other factors controlling gene-for-gene specificity (Luck et al. 2000). An important class of *R* genes includes putative cytoplasmic proteins that share a TIR domain. An example of this type of R protein is the well-characterized N protein of tobacco, which belongs to the TIR-NBS-LRR class of plant *R* genes and confers resistance to tobacco mosaic virus (TMV) (Whitham et al. 1994). Tobacco cultivars lacking a functional *N* gene do not show any resistance against TMV (i.e., HR), and loss-of-function *N* alleles such as the TIR deletion result in sensitive plants and disease (Durner et al. 1997; Dinesh-Kumar et al. 2000).

Following the initial perception, signals may be transduced through G-proteins, ion fluxes, reactive oxygen species, and/or phosphorylation cascades involving various kinases/phosphatases, among them mitogen-activated protein kinases (MAP kinases) (Hirt 1997). The activation of MAPKs is one of the earliest responses in plants challenged by avirulent pathogens or cells treated with pathogen-derived elicitors as shown by several recent gain-of-function studies (Zhang and Klessig 2001). For example, expression of a constitutively active MAPK kinase, *NtMEK2*, in tobacco induces the expression of defense genes and hypersensitive response-like cell death, which are preceded by the activation of two endogenous MAPKs, salicylic acid-induced protein kinase (SIPK) and wounding-induced protein kinase (WIPK) (Romeis et al. 1999; Klessig et al. 2000; Zhang and Liu 2001). Interestingly, MAPK activation is regulated at multiple levels. The dramatic rise in WIPK activity after TMV infection or elicitin treatment requires both increases in WIPK transcription and translation, as well as posttranslational phosphorylation. By contrast, activation of SIPK, like that for MAP kinases in yeast and animals, is regulated strictly at the posttranslational level by dual phosphorylation of threonine and tyrosine residues. We may assume that MAPK, just like in animals, play multiple roles in regulating HR-associated cell death (Hirt 1997).

HR is the best studied model for programmed cell death (PCD) in plants. Several biochemical and morphological parameters have been described for various types of cell death, and most of them can be observed during HR (Pontier et al. 1999). Characteristic features are cell shrinkage, cytoplasmatic condensation, chromatin condensation, DNA fragmentation and activation of various (cysteine) proteases (Pennell and Lamb 1997). In addition, the LOX-dependent peroxidative pathway, responsible for tissue necrosis, appears to be one of the features of HR programmed cell death, at least in tobacco (Rusterucci et al. 1999). Mitochondria play important roles in animal apoptosis and are implicated in salicylic acid (SA)-induced plant resistance to viral pathogens (Govrin and Levine 2000; Xie and Chen 2000).

A prominent feature of the HR is a massive increase in the generation of reactive oxygen species (oxidative burst), which both precede and accompany lesion-associated host cell death. Reactive oxygen species (ROS) have been implicated in an increasing number of physiological and pathological processes in all organisms. In plants, the production of ROS plays a major role during photooxidative conditions resulting from excess UV, ozone or light stress, or from other abiotic stresses like metal contamination or cold stress (Sandermann 2000). The deleterious effects of ROS are well known. However, superoxide radicals (O_2^-), hydrogen peroxide (H_2O_2) and hydroxyl radicals (OH·) are thought to play key roles in defense responses of plants against pathogens. Following infection, plants resistant to the invading pathogen develop a sustained increase of ROS (oxidative burst) (Baker and Orlandi 1995; Doke 1997). In a manner analogous to their participation in macrophage or neutrophil action, these ROS might be involved in directly killing invading pathogens. It should be noted, though, that neither O_2^- nor H_2O_2 are especially good candidates for killing bacteria or fungi directly. In addition, increases in H_2O_2 have been shown to induce the cross-linking of cell wall proteins and to enhance the peroxidase-catalyzed synthesis of lignin, thereby creating a physical barrier against pathogens (Brisson et al. 1994).

After a long debate regarding the source of ROS during HR, it seems to be clear that at least in tobacco and *Arabidopsis* a NADPH oxidase (respiratory burst oxidase homologues, rboh) generates massive amounts of superoxide (Torres et al. 1998, 2002). We might assume that many other plant species (e.g., parsley, tomato) carry homologues of these oxidases (Xing et al. 1997; Scheel 1998). On the other hand, plants seem to have several possibilities to produce ROS. Comparative biochemistry of the oxidative burst produced by rose and French bean cells revealed two distinct mechanisms and suggested that bean plants produce ROS via peroxidases (Bolwell et al. 1998).

An important feature that may be triggered by ROS is the induction of cell death in early plant defense responses. Treatment of soybean cells with high concentrations of H_2O_2 was shown to cause cell death (Shirasu et al. 1997). In contrast, another study has suggested that O_2^-, but not H_2O_2, is crucial for execution of cell death. In the *Arabidopsis lsd1* mutant elevated levels of O_2^-, but not H_2O_2, were able to induce spontaneous lesion formation (Jabs et al. 1996). However, despite repeated suggestions that ROS are involved in the signalling pathways leading to apoptosis and/or programmed cell death in animals, there still is no conclusive evidence that ROS are sufficient for the execution of cell death. Indeed, it has recently been suggested that ROS may be associated with, but not directly responsible for, apoptosis in animal cells (Jacobson 1996). Recent evidence points to the participation of nitric oxide (NO) in initiation and/or control of HR and to a combined action of NO, ROS, SA and ethylene (Van Camp et al. 1998; Torres et al. 2002; see also below, Sect. 5.3).

ROS can also serve as second messengers for the activation of defense gene expression. For example, elevated ROS levels induce the genes for glutathione-S-transferase, glutathione peroxidase and polyubiquitin, as well as peroxidases, catalases and other enzymes involved in ROS scavenging, chilling tolerance and pathogen resistance (Levine et al. 1994; Mehdy et al. 1996; Van Breusegem et al. 2001). Currently, the mechanism(s) by which redox signalling activates these genes is a matter of debate. The ability of ROS, and thus the cellular redox state, to activate plant defenses may parallel the mechanism by which oxidative stress induces the genes associated with animal immune and inflammatory responses. Activity of at least two defense-associated transcription factors, NF-κB and AP-1, have been shown to be regulated by active NADPH oxidase and the cellular redox state, respectively (Baeuerle and Baltimore, 1996; Fan et al. 2001). It is believed that many key events during initiation of HR and maintenance of local resistance are (at least partially) controlled by ROS.

5.1.2 Systemic Acquired Resistance (SAR)

Plants are defended against pathogens by constitutive and inducible barriers. Induced resistance may be expressed locally at the site of infection as well as systemically. The potential of plants to react to an invader by triggering local and systemic defense responses was first described by phytopathologists such as Carbone and Arnaudi (1930), Chester (1933) and Gäumann (1946). Later, using the model plant tobacco, induced resistance was shown in the uninfected upper leaves of plants, which had developed necrotic lesions on the lower leaves after inoculation with tobacco mosaic virus (TMV) (Ross 1966). This resistance was referred to as systemic acquired resistance (SAR). Pathogen-induced SAR has been observed in many species of di- and in monocotyledonous plants, and a large number of studies document how SAR can broadly protect plants encompassing the boundaries set by race-specific responses (Sticher et al. 1997). SAR is initiated as a consequence of a localized infection. Recognition of pathogens by elicitors released at the site of infection that interact with corresponding receptors is rapidly followed by modifications in ion fluxes, reactive oxygen species and phosphorylation events (see Sect. 5.1.1; Dangl and Jones 2001). This is followed by the activation of a signalling network that leads to a multiplicity of transcriptional events involved in various aspects of local and systemic resistance responses. A straightforward explanation for the systemic resistance is the production of a signal released from the infected leaf and translocated to other parts of the plant where it induces defense reactions. Interestingly, besides localized infection by pathogens, other treatments can also lead to systemic induced defenses in plants. For instance, colonization of roots with non-pathogenic bacteria can induce resistance in leaves (Pieterse and van Loon 1999). This phenomenon was termed induced systemic resistance. Also, viral infection can lead to the systemic induction of posttranscriptional gene silencing at sites distal from cells initially infected (Waterhouse et al. 2001). Environmental factors such as light or UV irradiation can also have an impact on SAR (Islam et al. 1998;

Mercier and Lindow 2001; Genoud et al. 2002). SAR expressed against a broad spectrum of pathogens caught the attention of many researchers, with its possible applications for novel approaches to plant protection. Thus, investigations on the molecular responses in infected and uninfected parts of induced plants became an important target for many groups. This led to the discovery of a subset of proteins induced locally but also systemically, termed PR- or pathogenesis-related proteins (Van Loon and Van Strien 1999). Many of these PRs were shown to have chitinase, 1,3-glucanase, lysozyme and other antibacterial or antifungal activities. With the development of genome-wide analyses, the number of defense-related proteins increased even more (Maleck et al. 2000; Schenk et al. 2000). The attention also became focused on endogenous regulators of defense responses such as salicylic acid (SA). Ectopic applications of SA were first found to induce PRs in tobacco and to protect the plant against TMV (White 1979). Later, SA was shown to be produced by plants after pathogen infection, locally and to a lesser extent systemically, making SA a possible signal for SAR (Sticher et al. 1997). A large number of studies were then oriented to understanding the role of SA in SAR. The following two sections will mainly focus on SA and its role in SAR to pathogens.

5.2 Salicylic Acid (SA) as a Signal for SAR

5.2.1 SA and Its Biological Relevance for SAR

SA is broadly distributed among many species and is a regulator for physiological processes such as thermogenesis or defense against pathogens (Raskin 1992). Observations on increases in the SA content of leaves and phloem cells after a localized pathogen infection, as well as the protective effect obtained when plants are treated with SA, have led to the hypothesis that SA might be a signal in SAR. The most compelling experiments in support of the importance of SA for SAR came from studies with mutants and transgenic plants that exhibit altered levels of SA (Table 5.1). In general, plants with low endogenous SA levels cannot mount efficient local and systemic defense responses to pathogens. On the other hand a number of mutants were found that accumulated constitutively high levels of SA and exhibited increased tolerance to pathogens (Table 5.1). The picture emerged that SA is a regulatory component for the induction of resistance and changes in its concentration activate of reactions leading to defense.

Since SA is required for induced defense, it was reasonable to determine if it is also the translocated systemic signal that moves from the lower to the upper leaves. Grafting and leaf excision experiments in tobacco and cucumber indicate that SA is a necessary component for the induction of local and

Table 5.1. Mutants and transgenic plant lines with altered levels of SA and compromised defense reactions

Species	Mutant/ transgene*	SA	Resistance			Reference
			P. parasitica	P. syringae	Other	
A. thaliana	acd2	+		+		Greenberg et al. (1994)
A. thaliana	acd5	+		–		Greenberg et al. (2000)
A. thaliana	agd2	+	+	+		Rate et al. (2001)
A. thaliana	lsd6	+	+	+		Weymann et al. (1995)
A. thaliana	lsd7	+	+			Weymann et al. (1995)
A. thaliana	cpr1	+	+	+		Bowling et al. (1994)
A. thaliana	cpr5	+	+	+		Bowling et al. (1997)
A. thaliana	cpr6	+	+	+		Clarke et al. (1998)
A. thaliana	cpr22	+				Yoshioka et al. (2001)
A. thaliana	dnd1	+	+	+		Yu et al. (1998)
A. thaliana	ssi1	+	–	+		Shah et al. (1999)
A. thaliana	mpk4	+	–	+		Petersen et al. (2000)
A. thaliana	eds5	–	–	–		Dewdney et al. (2000); Nawrath et al. (2002)
A. thaliana	sid2	–	–	–		Dewdney et al. (2000); Nawrath et al. (2002)
A. thaliana	NahG*	–	–	–		Nawrath and Métraux (1999)
A. thaliana	SA synthase*	+	+	+		Delaney et al. (1994)
N. tabacum	NahG*	–			TMV	Gaffney et al. (1993); Mauch et al. (2001)

systemic resistance, but that SA is not the primary mobile signal exported from the infected leaf to other parts of the plant (reviewed in Métraux 2001). However, it was shown in tobacco and cucumber that SA synthesized after inoculation can be transported from the infected leaf to the upper leaves by the phloem before resistance was detectable (Shulaev et al. 1995; Mölders et al. 1996). Possibly, SA produced in high amounts at infection sites can be translocated along with another primary mobile signal and induce resistance in upper leaves. In tobacco, volatile methyl salicylate (MeSA) is produced from SA after infection. MeSA can induce defense reactions by conversion to SA (Shulaev et al. 1997; Seskar et al. 1998) and was proposed to be an additive to SA for signalling within a plant and to act as a signal for communication between plants. Thus, while SA is necessary for local and systemic induction of resistance responses, it is likely that SA as well as another systemic signal could be involved in SAR. The nature of such a systemic signal remains unknown at this time.

For the sake of completeness, we will briefly review other endogenous molecules that have been identified as signals involved in SA-independent activation of induced resistance. These compounds include ethylene and octadecanoic acid derivatives such as jasmonic acid (JA), methyl jasmonate (MeJA) and 12-oxo-phytodienoic acid (OPDA).

JA mediates the resistance of plants to pathogens and insects (Tahler et al. 2001) or insect communities (Thaler et al. 1999). As for SA, applications of JA to plants can lead to protection, and JA was shown to accumulate after wounding inflicted by chewing insects or necrotrophic pathogens (Creelman and Mullet 1997). *Arapidopsis* plants impaired in JA accumulation such as the *fad3/7/8* (fatty acid desaturase) or the JA-insensitive mutant *jar-1* (jasmonic acid resistant) are more susceptible to necrotrophic pathogens such as *Pythium* (Staswick et al. 1998; Vijayan et al. 1998) and show decreased expression of the plant defensin gene *PDF1.2*. The *cor-1* (coronatine resistant) mutant is insensitive to MeJA and more susceptible to the necrotrophic pathogen *Botrytis cinerea*, while its resistance against biotrophs such as *Pseudomonas* or *Peronospora* is unaltered (Feys et al. 1994; Thomma et al. 1998). A complex regulatory network exists within the jasmonate family. For example, evidence was presented that intermediates in the JA biosynthesis such as OPDA can also act as signals. Mutants of OPDA reductase (*opd3*) can accumulate OPDA, but not JA, and still induce *PDF1.2* expression after infection, while retaining resistance to the insect *Bradysia impatiens* and the fungus *Alternaria brassicicola* (Stintzi et al. 2001). *Arabidopsis* plants overexpressing S-adenosylmethyl-L-methionine-jasmonic acid carboxylmethyltransferase (JMT) contain elevated levels of MeJA, while JA remains low. Such plants exhibit elevated levels of *PDF1.2* and increased resistance to *A. brassicicola* (Seo et al. 2001).

Infection or treatment with elicitors can increase the endogenous levels of ethylene in plants and induce certain PRs (Boller et al. 1983; Potter et al. 1993). The ethylene-insensitive *Arabidopsis* mutant *ein2* has an increased susceptibility to an avirulent isolate of *B. cinerea* (Thomma et al. 1998) and in soybean, ethylene-insensitivity impairs resistance to avirulent strains of *Phythophthora megasperma* (Hoffman et al. 1999). A similar situation exists in tobacco, where an ethylene-insensitive mutant lacks resistance to the nonhost pathogen *Pythium sylvaticum* (Knoester et al. 1998). In tomato challenged with virulent bacterial (*Xanthomonas campestris* pv *vesicatoria* and *Pseudomonas syringae* pv *tomato*) and fungal (*Fusarium oxysporum* f. sp *lycopersici*) pathogens, disease symptoms were reduced in mutants affected in ethylene synthesis and perception. PR-1B1 mRNA accumulation in response to *X. c. vesicatoria* infection was not affected by ethylene insensitivity, indicating that ethylene is not required for defense gene induction (Lund et al. 1998). Similarly, soybean mutants with reduced sensitivity to ethylene develop fewer symptoms to virulent strains of *P. syringae* or *Phytophthora sojae* (Hoffman et al. 1999). The ethylene-response factor 1 (ERF1) is a regulator of the ethylene responses after pathogen attack in *Arabidopsis*. The ERF1 transcript is induced on infection by *B. cinerea*, and overexpression of ERF1 in *Arabidopsis* is sufficient to confer resistance to necrotrophic fungi such as *B. cinerea* and *Plectosphaerella cucumerina* (Berrocal-Lobo et al. 2002). It was proposed that ethylene might possibly control the development of symptoms by controlling the expression of the HR (Knoester et al. 1998; Lund et al. 1998; Ciardi et al. 2001). A possible explanation could be that, besides its role in defense during incompatible interactions, HR associated events such as ROS generation might help plants during compatible plant pathogen interactions. This argument was also used to explain increased resistance to virulent strains of *P. syringae* in *Arabidopsis* mutants impaired in the HR response (Stone et al. 2000).

A number of studies have shown that signals can also act in combination. For instance, after inoculation with necrotrophic fungi such as *A. brassicicola*, JA levels increase locally and systemically and act together with ethylene to induce the expression of *PDF1.2*, a plant defensin gene (Thomma et al. 1998). Induction of the *HEL*, *CHIB* and *PDF1.2* gene in *Arabidopsis* infected with *Erwinia carotovora* also depends on an interaction between the ethylene and JA pathways (Norman-Setterblad et al. 2000). Regulators leading to the combined induction of the ethylene and JA pathways have already been identified. For example, the *Arabidopsis cev1* mutant has constitutively activated ethylene and JA pathways and is resistant to various powdery mildew isolates (Ellis and Turner 2001). The CEV1 protein seems to be at a node where two signalling pathways converge. Recent data on *Arabidopsis* have shown that full expression of the pathogen-induced glutathione-*S*-transferase gene *AtGSTF2* and to a lesser extent At*GSTF6* is the product of the combined activity of SA and ethylene signalling. At*GSTF2* is likely to reduce the consequences of oxidative damage caused by pathogens or by the hypersensitive reaction (Lieberherr et al. 2003). A positive cooperation between the ethylene and SA pathways was also observed in the response of *Arabidopsis* to *P. cucumerina*, whereas ethylene and SA negatively cooperate in response to infection by *P. syringae* pv. *tomato* DC3000 (Berrocal-Lobo et al. 2002).

In tomato, abscisic acid determines basal susceptibility to *B. cinerea* and suppresses the SA-dependent defense pathway (Audenaert et al. 2002). The latter example illustrates that while the same signals can be observed in a diversity of plants, their intermodulation might not be identical in all species.

Several studies also indicate that certain signalling pathways can be mutually inhibitory. For example, JA-dependent expression of *PDF1.2* is inhibited by SA (Penninckx et al. 1998; Dewdney et al. 2000; Gupta et al. 2000). Studies with the *cev1* mutant also show a mutual antagonism between the ethylene and the JA pathways (Ellis and Turner 2001). The *Arabidopsis* SSI1 protein was proposed to connect both JA and SA signalling pathways (Shah et al. 1999). It is now interesting to know how SSI1 is regulated. It was proposed that the SA-dependent defense responses were directed at biotrophic pathogens such as *Peronospora* or *Pseudomonas*, whereas JA and ethylene-dependent responses operated mostly against necrotrophic organisms (Thomma et al. 2001) or insects (Bostock et al. 2001).

Interestingly, the trade-off between SA- and JA-mediated signalling could be documented in greenhouse and field experiments with tomato. Treatments with the SA mimic, benzothiadiazole (BT; BION), attenuate the JA-induced expression of the antiherbivore defense-related enzyme polyphenol oxidase and compromise host-plant resistance to larvae of the beet armyworm, *Spodoptera exigua*. Conversely, JA treatment reduces the expression of *PR* genes induced by BT and partially reverses the protective effect of BT against *P. syringae* pv. *tomato* (Thaler et al. 1999). In tobacco, an inverse cross-talk between the effects of SA and JA-mediated signalling was also observed. SAR to tobacco mosaic virus (TMV) is impaired when gene silencing reduced the expression of the phenylalanine ammonialyase (PAL) gene, whereas overexpression of PAL enhances SAR. Plants with reduced SAR exhibited more resistance to larvae of *Heliothis virescens*, while larval resistance was decreased in plants with elevated phenylpropanoid levels (Felton et al. 1999). Thus, an effective utilization of induced plant resistance under agricultural conditions where a multiplicity of pests coexist will require a good understanding of potential signalling conflicts in the defense responses of the plant (Heil and Baldwin 2002).

Finally, a number of studies in *Arabidopsis* clearly indicate the existence of signalling pathways that are independent of SA, JA or ethylene. For example, in *Arabidopsis*, *PAD2*-mediated resistance to *Phytophthora porri* or the *RPP7-*, *RPP8- and RPP13*-mediated resistance to *Peronospora parasitica* function independently of these signals (McDowell et al. 2000; Bittner-Eddy and Beynon 2001; Roetschi et al. 2001).

5.2.2 Biosynthesis and Metabolism of SA

Given the implication of SA in the induction of resistance, renewed attention has been given to its biosynthesis. Early studies have shown that SA derives from the shikimate-phenylpropanoid pathway (reviewed in Sticher et al. 1997). Two routes from phenylalanine to SA have been described that differ at the hydroxylation of the aromatic ring. Cinnamic acid (CA), the product of the conversion of phenylalanine (Phe) by phenylalanine-ammonia lyase (PAL), either can be hydroxylated to form *ortho*-coumaric acid followed by oxidation of the side chain, or, alternatively, the side chain of CA is first oxidized to give benzoic acid (BA), which is then hydroxylated in the *ortho* position (reviewed in Sticher et al. 1997). In tobacco, SA was proposed to be synthesized from free BA (Yalpani et al. 1993), but more recent results indicated that benzoyl glucose, a conjugated form of BA is more likely to be the direct precursor of SA (Chong et al. 2001). In cucumber, potato and rice the synthesis of SA was also shown to derive from CA (reviewed in Sticher et al. 1997).

These studies point towards a biosynthetic pathway from phenylalanine via CA and BA, but the exclusive role of this route in pathogen-induced SA was never fully assessed.

Arabidopsis also produces SA locally and systemically after pathogen infection or treatment with UV (Summermatter et al. 1995; Nawrath and Métraux 1999; Nawrath et al. 2002). Support for a Phe and CA-derived biosynthetic pathway in *Arabidopsis* was provided using 2-aminoindan-2-phosphonic acid (AIP), an inhibitor of PAL. AIP-treated plants had lower amounts of SA and diminished resistance to *P. parasitica*, supporting a major role for CA-derived SA in SA synthesis and resistance (Mauch-Mani and Slusarenko 1996). Contrasting results were obtained in *Arabidopsis* after the characterization of the SA-induction deficient mutant *sid2*. The *sid2* mutation was mapped to a gene *(ICS1)* encoding isochorismate synthase (ICS) (Wildermuth et al. 2001). *ICS1* includes the highly conserved chorismate-binding domain and shares 57 % amino acid sequence identity with a *Catharanthus roseus* ICS and 20 % amino acid identity with the ICS from bacteria that have confirmed biochemical activities (Serino et al. 1995; Wildermuth et al. 2001). *ICS1* is induced locally and systemically upon a localized pathogen infection and the level of SA after infection in *sid2* mutants is only 5–10 % of the wild type levels, while resistance to fungal or bacterial pathogens is reduced (Nawrath and Métraux 1999; Wildermuth et al. 2001). These results provide solid support for an *ICS*-mediated SA synthesis required for SAR. Studies using PAL inhibitors in *Arabidopsis* have to be reevaluated now in light of these results. Besides the strong homology to *Catharanthus ICS*, ESTs for *ICS* have been annotated in soybean and tomato (Wildermuth et al. 2001). Thus, it is likely that higher plants produce pathogen-induced SA from isochorismate, a biosynthetic pathway typical for bacteria. Plastid-localized synthesis of SA is indicated by the presence of a plastid transit peptide and cleavage site in the *ICS1* gene, making it likely that the SA pathway in *Arabidopsis* might derive evolutionarily from prokaryotic endosymbionts (Wildermuth et al. 2001). The promoter of *ICS1* contains W-box elements. Such elements are recognized by various WRKY transcription factors and are generally involved in the regulation of pathogen or stress responses (Eulgem et al. 2000). The *ICS1* promoter also comprises a binding site for Myb transcription factors that regulate genes for plant defense and associated secondary metabolism (Yang and Klessig 1996; Bender and Fink 1998). BZip or NF-κB motifs typically required for the induction of *PR1* by SA were not found in the *ICS1* gene (Cao et al. 1997; Ryals et al. 1997; Wildermuth et al. 2001). Indeed, *ICS1* expression is independent of SA since wild type expression levels are observed in infected SA-depleted NahG plants (Wildermuth et al. 2001). *ICS1* expression and attending defense responses are likely to be under the control of another (earlier) signal than SA.

The plastidic localization of SA synthesis implies a release of SA to its presumed site of action in the cytoplasm. Interestingly, the mapping of another SA-induction deficient mutant, *eds5/sid2*, identified a membrane protein

(Nawrath et al. 2002). EDS5/SID1 is homologous to the bacterial multidrug and toxin extrusion proteins (MATE) that have recently been reported to occur in *Arabidopsis* (Brown et al. 1999; Debeaujon et al. 2001; Diener et al. 2001). Preliminary results suggest that EDS5/SID1 is localized at the chloroplast envelope (Nawrath et al. unpubl. results), and it will now be very interesting to learn more about the nature of the substrate(s) transported by EDS5/SID1.

Since the isochorismate pathway might not be unique for *Arabidopsis* (Wildermuth et al. 2001), the relative importance for the induction of LAR and SAR of CA-derived and ICS-derived SA needs now to be evaluated in other plants. If both the CA and the isochorismate pathways operate in a same species, each pathway might possibly be induced by specific stimuli. Results obtained with the *Arabidopsis sid2* mutant do not support this, since widely different stimuli such as virulent or avirulent pathogens, ozone stress or callus formation are ineffective in the *sid2* mutant (Nawrath and Métraux 1999). Alternatively, the CA pathway might only contribute to basal levels of SA present in uninfected plants. Besides CA, another source of the basal levels of SA was proposed to result from the action of a second *ICS* gene (*ICS2*), the transcripts of which remain undetected in infected or uninfected leaves of *Arabidopsis* (Wildermuth et al. 2001). More experiments will help to clarify the function and regulation of CA- and ICS-derived SA in *Arabidopsis*, but also in those species where CA was proposed as a main precursor for pathogen-induced SA.

5.2.3 The Activation of Defense Mechanisms by SA

An understanding of the molecular action of SA implies knowledge of potential receptors of SA. Binding of SA with its receptor is thought to activate downstream reactions such as the activation of specific genes.

Catalase was initially proposed as a receptor for SA (Chen et al. 1993). Binding and inactivation of catalase would result in an increase of H_2O_2 that could activate defense gene expression or act as an antimicrobial barrier at the site of invasion (Chen et al. 1993). However, the catalase inhibition hypothesis was questioned on various grounds by several authors (reviewed in Mauch-Mani and Métraux 1998). Another action of SA was proposed to affect the redox status of the cells. It is based on the ability of SA to form free radicals upon inhibition of heme-containing enzymes such as peroxidase or catalase (Durner and Klessig 1995, 1996). Phenolic free radicals can be potent initiators of lipid peroxidation, the products of which might activate defense reactions (Farmer et al. 1998). It would now be interesting to see whether SA binding to peroxidase or catalase leads to the formation of active lipid peroxides in sufficient amounts at the right time and place to induce the defense response. Besides catalase and peroxidase SA was also found to bind to another protein. The affinity of this protein for SA was about 150-fold higher than that of SA for catalase. The ability of functional analogues to compete with the binding of labeled SA to this protein correlated with their ability to induce resistance and PR1 gene expression (Du and Klessig 1997). The biological importance of this SA-binding protein now remains to be determined.

Other studies concerning the action of SA were directed at its role in the signal transduction cascade from pathogen perception to defense. Pathogens, cell wall-derived- and pathogen-derived elicitors, reactive oxygen species (ROS), SA, as well as biologically active analogs of SA were reported to induce the activity of protein kinases belonging to the MAP kinase (MAPK) family (Romeis et al. 1999; Cardinale et al. 2000; Nuhse et al. 2000; Nürnberger and Scheel 2001; Zhang and Klessig 2001). Signalling cascades that involve the sequence of reversible phosphorylations of MAPK by MAPK kinases (MAPKK), themselves phosphorylated by MAPKK kinases (MAPKKK), are well known in eukaryotic signal transduction from external stimuli to cellular responses (Romeis et al. 1999; Nürnberger and Scheel 2001; Wraczek and Hirt 2001; Zhang and Klessig, 2001). Earlier studies in tobacco showed that SA and biologically active analogs could induce a protein kinase belonging to the MAPK family (Zhang and Klessig 1997). The involvement of the MAPK pathway in the activation of disease resistance in tobacco was further supported by studies showing that the MAPKK NtMEK2 activates SIPK. This is followed by a HR-like cell death and activation of the expression of 3-hydroxy-3-methylglutaryl CoA reductase (HMGR) and L-phenylalanine ammonia-lyase (PAL), two defense genes encoding key enzymes of the biosynthetic pathway of phenolics related to defense (Yang et al. 2001). Interestingly, SA was not found to be involved in the NtMEK2-mediated activation of HR (Yang et al. 2001), indicating the existence of alternative signalling cascades for ROS and SA. Similarly, the existence of different MAPK cascades was also inferred from the study of the flagellin cascade in *Arabidopsis* (Asai et al. 2002). In *Arabidopsis*, H_2O_2 activates the MAPKKK ANP1 that can activate the SIPK analogs AtMPK3 and AtMPK6 apparently without the implication of SA (Kovtun et al. 2000). MAPKs were described that negatively regulate defense reactions. Inactivation of the *Arabidopsis* AtMAPK4 by transposon insertion leads to plants with *NPR1*-independent resistance to virulent oomycete and bacterial pathogens, constitutive PRs and elevated levels of SA without spontaneous necrotic lesions (Petersen et al. 2000). In addition, AtMAPK4 is activated rapidly after wounding and regulates positively the jasmonate-dependent signalling pathway. Another mutation in a putative MAPKKK gene in *Arabidopsis* leads to a phenotype characterized by increased resistance to powdery mildew and *Pseudomonas syringae* (Frye et al. 2001). This *edr1* mutant (enhanced disease resistance) is dependent on SA and NPR1, without effect on the accumulation of SA and PR1. This suggests that EDR1 acts at the top of a MAP kinase cascade that negatively regulates SA-inducible defense responses. Summarizing, while the activation of MAPKs by SA (presumably preformed) has been reported in some instances, many studies seem to show that kinase cascades operate without a direct implication of SA. Such signalling cascades would precede downstream defense responses, some of which are SA-dependent.

Studies with *Arabidopsis* impaired in various aspects of induced resistance lead to the positioning of the action of SA downstream of EDS1 and PAD4 and upstream of the ankyrin-repeat protein NPR1 (Glazebrook 2001). In response to SA, NPR1 localizes in the nucleus where it can interact with TGA transcription factors of the bZIP protein family and induce *PR1* transcription (Lebel et al. 1998; Zhang et al. 1999; Despres et al. 2000; Kinkema et al. 2000; Zhou et al. 2000). The nuclear protein SNI1 negatively regulates SA-induced *PR* gene expression, presumably by interfering directly with a specific DNA sequence or via a transcription factor (Li et al. 1999). Diverse ligands activating orphan receptors in animals [in particular the peroxisome proliferator-activated receptor (PPAR) subfamily], such as flufenamic acid, tolbutamide, indomethacin and clofibrate can induce programmed cell death in soybean cell-suspension cultures. The PPAR ligands as well as SA were found to differentially induce novel cDNAs of hitherto unknown function that might be linked to HR, but not SAR (Tenhaken et al. 2001).

The molecular action of SA was found to be distinct in relation to virus resistance. SA can interfere specifically with virus replication and movement (Murphy et al. 2001). The plant defense signalling downstream of SA is likely to be separated in a fungus- and bacteria-specific and in a virus-specific branch (Murphy et al. 2001; Murphy and Carr 2002). SA treatment inhibits replication or movement of several positive-sense RNA plant viruses. In tobacco and *Arabidopsis*, nonlethal concentrations of cyanide (CN) and antimycin A (AA) have similar effects without triggering induction of the *PR-1* genes. Besides virus resistance induced by CN, AA or SA, the mitochondrial alternative oxidase is also induced independently of *NPR1*, a gene that plays a key role downstream of SA in the induction of PRs in *Arabidopsis* (Wong et al. 2002). This shows that in *Arabidopsis* and in tobacco, resistance to viruses can be induced via a distinct branch of the defensive signal transduction pathway. AA-, SA- and ROS-induced expression of a mitochondrial alternative oxidase gene was also observed in tobacco cell cultures, and the mitochondrion was proposed to function in conveying stress signals to the plant nucleus (Maxwell et al. 2002).

A possible action of SA was also considered in relation to conditioning. Also referred to as priming, this process implies that SAR-derived signals provide the ability to induce defense reactions more strongly and rapidly after infection (Conrath et al. 2002). This might in fact be the situation experienced by upper leaves of plants infected locally on lower leaves. The arrival of a putative systemic signal might condition, prime or potentiate the tissue to a stronger response after an attack by a pathogen. Priming has been demonstrated in cell cultures pretreated with SA or functionally related inducers prior to elicitor exposure. Pretreatments with SA, 2,6-dichloroisonicotinic acid (INA) or benzothiadiazole (BTH; BION) potentiate the elicitor-induced expression of defense-related phenylpropanoid genes (Katz et al. 1998; Thulke and Conrath 1998). Observations were also made in cucumber hypocotyls, where pretreatments of the seedlings with INA or BTH lead to increased competence for elicitor-induced H_2O_2 generation (Kauss et al. 1999). Priming has also been observed in whole plants. *Arabidopsis* pretreated with pathogens or BTH show an increase in the sensitivity to *P. syringae*-induced activation of the *PAL* gene and callose deposition, two reactions that are not induced by BTH alone. Priming by BTH and pathogen infection for resistance to *P. syringae* requires the activity of the *NIM1/NPR1* gene (Kohler et al. 2002). The non-protein amino acid β-aminobutyric acid (BABA) protects *Arabidopsis* from infection with *Peronospora parasitica*. BABA acts by potentiating the tissue to a stronger deposition of callose-containing papillae at the fungal infection sites. In response to infection with virulent *P. syrinage*, the effect of BABA manifests itself by a potentiation of the induction of *PR1* (Zimmerli et al. 2000). Interestingly, the effect of BABA against *P. parasitica* is independent of the SA, JA and ethylene signalling pathways, whereas BABA potentiation to *P. syringae* is dependent on operational SA signalling (Zimmerli et al. 2000). Future experiments should now be aimed at the action of SA or of the putative systemic signal in conditioning of defenses in systemic leaves.

5.3 The Oxidative Burst and Nitric Oxide (NO)

Because of the diversity of its physiological functions and general ubiquity, NO has attracted a great deal of attention. NO first captured the interest of biologists when this highly reactive radical was identified as the endothelium-derived relaxing factor, a potent endogenous vasodilator. In 1992 NO was recognized by *Science* magazine as Molecule of the Year. Further investigations led to the finding that NO is a multifunctional effector involved in numerous mammalian (patho)physiological processes, including relaxation of smooth muscle, inhibition of platelet aggregation, neural communication and immune regulation. The Nobel Price for Medicine in 1998 went to three pioneers of NO research. NO is now recognized to be an intra- and intercellular mediator of animal cell functions, and the very reactions of NO with redox centers in proteins and membranes that were originally identified with injurious or polluting effects of this sometimes toxic molecule are now being established as molecular components of many cellular signal transductions (Mayer and Hemmens 1997; Lamas et al. 1998; Wink and Mitchell, 1998).

Until recently, research on the effects of NO in plants focused on atmospheric pollution by the oxides of nitrogen, NO and NO_2 (nitrate). Uptake of

NO into foliage, as well as its subsequent metabolism and phytotoxicity, are well documented (Wellburn 1990; Hufton et al. 1996). Later it became clear that plants not only *respond* to atmospheric NO, but they also are able to *produce* substantial amounts of NO (Wildt et al. 1997). Mounting evidence also suggests that NO is a novel effector of plant growth, development and defense (Durner and Klessig 1999; Durner et al. 1999; Wendehenne et al. 2001). For example, NO was shown to induce leaf expansion, root growth and phytoalexin production (Leshem 1996; Noritake et al. 1996). In the following paragraphs we will briefly summarize our current view of NO signalling in plants, focusing on the recent findings during the past year on NO's role in plant defense responses against pathogens. In addition, we will briefly outline the interaction and cross-talk of NO with other plant signalling molecules such as ethylene and SA.

5.3.1 Biological Chemistry and Properties of NO

Understanding the biological chemistry and properties of NO is fundamental to integrate its various biological effects. NO chemistry implicates an interplay among the three redox-related species nitric oxide radical (NO), nitrosonium cation (NO^+) and nitroxyl anion (NO^-). In biological systems, NO reacts rapidly with molecular oxygen (O_2), superoxide anion (O_2^-) and transition metals. NO's interplay with reactive oxygen species is summarized in Chapter 4.8.2. The reaction of NO with O_2 results in the generation of higher nitrogen oxides (including NO_2, N_2O_3 and N_2O_4), which can either react with cellular amines and thiols or simply hydrolyze to form nitrite (NO_2^-) and nitrate (NO_3^-) (Halliwell and Gutteridge 1989; Wink and Mitchell 1998; Wendehenne et al. 2001). NO_2^- and NO_3^- are assumed to be NO's final metabolite. NO also forms complexes with the transition metals found in heme- or iron cluster-containing proteins, thus forming iron-nitrosyl complexes (Stamler et al. 1992; Stamler 1994). This process alters the structure and functions of the target proteins. In vertebrates, soluble guanylate cyclase is a well-known enzyme physiologically modified by NO. Conformational changes induced by NO binding at the heme active site of the protein lead to enzymatic conversion of guanosine triphosphate to guanosine cyclic 3',5'-monophosphate (cGMP), a cellular messenger involved in vasodilatation and smooth muscle relaxation. In contrast, when heme or iron clusters of metalloenzymes participate in the catalysis, formation of iron-nitrosyl complexes results in the inhibition of the corresponding enzyme activity as reported for cytochrome P450s, cytochrome c oxidase, aconitase or NOS itself (Wink and Mitchell 1998; Navarre et al. 2000). One electron oxidation of NO· leads to NO^+. This oxidation can be supported by Fe(III)-containing metalloproteins. The Fe(III)NO complex appears to undergo a charge transfer reaction to form Fe(II) NO^+, which subsequently may release NO^+ in an electron rich environment. NO^+ mediates electrophilic attack on reactive sulfur, oxygen, nitrogen and aromatic carbon centers, with thiols being the most reactive center (Mayer and Hemmens 1997). This chemical process is referred to as nitrosation. Nitrosation of sulfhydril centers (leading to S-nitrosothiols) of many enzymes/proteins has been described, and the resulting chemical modification, in many cases, affects activity. Some example are inhibition of the N-methyl-D-aspartate receptor, glyceraldehyde-3-phosphate dehydrogenase, caspase-3 or of the DNA repair protein G6-methylguanine-DNA-methyltransferase, and activation of the small GTP-binding p21ras, of L-type Ca^{2+} channel and of the Ca^{2+} channel ryanodine receptor (Brüne et al. 1998). Such modifications are reversible and protein S-nitrosylation/denitrosylation may represent an important transduction regulatory mechanism. In addition, a variety of S-nitrosothiols including S-nitrosoglutathione (GSNO) and S-nitrosocysteine (CySNO) may serve as NO "pools" and have been shown to confer NO-like biological activities such as antimicrobial effects, vasodilatation or platelet aggregation inhibition (Jia et al. 1996). These functions are usually attributed to NO release (NO^-, NO or NO^+ release), although there is some evidence that S-nitrosothiols can also exhibit direct effects (Mayer and Hemmens 1997).

Another important reaction of S-nitrosothiols is transnitrosation. Transnitrosation is defined as NO^+ transfer from a S-nitrosothiol to a thiol. This process is potentially complex and leads to the generation of a variety of species including NO, nitrous oxide (N_2O) and NO_2^- (Wink and Mitchell 1998). However, it may represent an important new mode of cellular signalling control since proteins including the N-methyl-D-aspartate receptor, protein kinase C, glutathione peroxidase and glyceraldehyde-3-phosphate dehydrogenase can all be modulated by transnitrosation (Mayer and Hemmens 1997).

One electron reduction of NO generates NO^-. NO^- is also formed biologically by nitric oxide synthases (NOS; see next section). The physiological meaning of NO^- has not been clarified. Some works suggest it may act as a stabilized form of NO (Mayer and Hemmens 1997). In addition, NO^- is believed to react with Fe(III) heme and to mediate sulfhydryl oxidation of target proteins (Wink and Mitchell 1998).

In sum, NO chemistry is complex and involves the three redox-related species of NO^+, NO^- and NO· (and derived products such as $ONOO^-$), each species undergoing distinctive properties and reactivities. NO appears to act primarily by activating guanylate cyclase. In addition, nitrosation of sulfhydryl centers seems a major cellular process to regulate protein functions. In order to simplify, in vertebrates NO signals are typically designed as either cGMP-dependent (pathways controlled by NO-mediated activation of guanylate cyclase) or cGMP independent (signalling by S-nitrosation).

5.3.2 Biosynthesis of NO in Plants

The presence and synthesis of NO in plant cells is undisputed. The question to be answered is *how* do plants produce NO? However, our understanding of NO production by plants is very limited. In animal cells, the biosynthesis of NO is primarily catalysed by the enzyme NOS (Lamas et al. 1998). Three NOS isoforms originally have been identified based on the tissue source: neuronal NOS (nNOS, type I), inducible NOS (iNOS, type II) and endothelial NOS (eNOS, type III) (Mayer and Hemmens 1997). All three NOS isoforms are highly related and show about 50 to 60 % identity in their amino acid sequence. Each NOS is bi-domain enzymes consisting of an N-terminal oxygenase and C-terminal reductase (Nathan and Shiloh 2000). The oxygenase domain contains a cytochrome P-450 type heme center and a binding site for the cofactor tetrahydrobiopterin (BH_4) (Ninnemann and Maier 1996). The reductase domain contains NADPH, FAD and FMN binding sites and exhibits significant homology (about 58 %) with NADPH cytochrome P-450 reductase. Both domains are connected by a calmodulin (CaM) binding site in the midle of the enzyme. In addition, each NOS has a different N-terminal extension, which determines the intracellular localization of the enzyme (Nathan and Xie 1994).

All NOS are active as homodimer. Dimerization occurs through the heme domains and binding of BH_4 and CaM stabilizes the dimer by promoting conformational changes. In their active form, NOS catalyses the conversion of L-arginine to L-citrulline and NO in a two-step reaction: oxygenation of L-arginine to *N*-hydroxyarginine, followed by oxygenation of *N*-hydroxyarginine to L-citrulline and NO (Hevel and Marletta 1994). The reaction requires O_2 and NADPH. The catalytic mechanism involves flavin-mediated electron transport from C-terminal bound NADPH to the heme iron, where O_2 is reduced. Bound CaM allows electron transfer from flavins to the heme group and thus functions as a molecular switch between the two domains of NOS.

Initial evidence for the presence of a mammalian-type NOS in plants was first reported in 1996 (Cueto et al. 1996; Leshem 1996; Ninnemann and Maier 1996). For example, NOS activity was found in roots and nodules of leguminous plants, where it was thought to be induced by *Rhizobium* lipopolysaccharides (Cueto et al. 1996). Perhaps, NO is involved in establishing symbiosis as well as in mediating nitrogen fixation through binding the heme group of redox sensors such as leghemoglobin or FixL (Mathieu et al. 1998). More recently, it has been shown that NO and possibly NOS play a prominent role in plant defense against microbial pathogens. Infection of resistant, but not susceptible, tobacco with tobacco mosaic virus resulted in enhanced NOS activity (Durner et al. 1998). Similar results were obtained with soybean cells and *Arabidopsis* by measuring both the activity of NOS and the subsequent release of NO in response to either a bacterial pathogen (*Pseudomonas syringae*) or an elicitor. Strikingly, NOS inhibitors were able to compromise the hypersen-

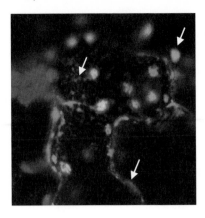

Fig. 5.1. (in color at end of book) In vivo imaging of a pathogen-induced NO burst in tobacco. NO production was visualized by confocal laser scanning microscopy of a cell treated with cryptogein, an elicitor from *Phythopthora* ssp. The cell was loaded with DAF-2DA, a specific fluorophore for NO. Six min after treatment with 50 nM cryptogein, DAF-2DA fluorescence was localized along the plasma membrane and to distinct regions within the chloroplasts as well as to other subcellular structures, which may represent peroxisomes (*arrowheads*). Experimental details were as described previously (Foissner et al. 2000)

sitive resistance response in *Arabidopsis* and the induction of programmed cell death in elicitor- or pathogen-treated soybean suspension cells (Delledonne et al. 1998). In vivo imaging of an elicitor-induced NO burst (using cryptogein from *Phythophthora infestans*) in tobacco also points to the involvement of a NOS-like enzyme during plant defense reactions (Fig. 5.1; Foissner et al. 2000). Furthermore, tobacco resisting infection by *Ralstonia solanacearum* exhibited elevated levels of NOS activity (Huang and Knopp 1998). Immunoblot analyses with antibodies made against rabbit brain NOS or mouse macrophage NOS were also found to react with tobacco and maize extracts that had high levels of NOS (Huang and Knopp 1998; Ribeiro et al. 1999). Electron microscopy immunolocalization using such antibodies showed the presence of NOS-like protein in the matrix of peroxisomes and chloroplasts in pea leaves. Moreover, immunofluorescence experiments in maize roots allowed researchers to localize a NOS-like protein in the cytosol of cells in the division zone and in the nucleus in the elongation zone (Ribeiro et al. 1999). Such growth phase-dependent subcellular localization of NOS is not reported in animal cells and could be specific to plants. However, it is yet to be rigorously established that the cross-reacting proteins correspond to plant NOSs. In this regard, it is disconcerting that the putative tobacco NOS has a M_r of 56 kDa, while the maize NOS was 166 kDa (mammalian NOSs range in size from 130–160 kDa).

The above findings would suggest the involvement of a NOS in plant defense responses. However, many questions remain. Although there is significant evidence that NO accumulation is needed to contain pathogen growth and spread, it is unclear whether NO (and/or its activated derivatives) are directly toxic to microbial pathogens in plants. Furthermore, it is not known whether NOS-like activity in plants originates from a mammalian-type NOS. Plant NOS activity and neuronal rat NOS differ in their degree of sensitivity to NOS inhibitors (Cueto et al. 1996; Delledonne et al. 1998; Durner et al. 1998).

The specificity of NOS inhibitors injected into plants needs to be elucidated as well. For example, it was suggested that inhibitors of NOS might inhibit polyamine oxidase-type enzymes, which can use L-arginine as a substrate to produce biologically relevant reactive oxygen species (ROS) (Allan and Fluhr 1997). In addition, it is unclear whether a plant can provide all cofactors necessary for NOS function. In mammals, NOS activity depends on the cofactor tetrahydro-L-biopterin (Bredt and Snyder 1990; Poulos et al. 1998). Although its precise function is still not known, this cofactor is essential for the coupling of NADPH-dependent O_2 activation to NO synthesis. To date, there is no information regarding whether or not tetrahydro-L-biopterin is a natural constituent of plants. Finally, a plant NOS gene has yet to be cloned, although the plant homolog of a protein that inhibits animal nNOS has been identified (Jaffrey and Snyder 1996).

Contrary to the common view, NO synthesis is not confined to organisms containing nitric oxide synthase (NOS). Rather, NO is also a byproduct of denitrification, nitrogen fixation or respiration. In most cases NO production has been linked to the accumulation of NO_2 in plant tissues (Klepper 1990, 1991). NO is produced non-enzymatically through light-mediated conversion of NO_2 by carotenoids (Cooney et al. 1994) or enzymatically from NO_2 by NADPH nitrate reductases (Wildt et al. 1997). NO emission from plants occurs under stress situations such as herbicide treatment or even under normal growing conditions, and it is believed that the ability to produce NO is common to many, if not all, plant species (Wildt et al. 1997). Recently, nitrate reductase (NR) has been established as a major source of NO in plants through one-electron reduction of nitrite using NAD(P)H as an electron donor (Yamasaki 2000). In sunflower, spinach or maize, NO production by leaf extracts or intact leaves was unaffected by NOS inhibitors. It was concluded that in non-elicited leaves NO is produced in variable quantities by NR depending on the total NR activity, the NR activation state and the cytosolic nitrite and nitrate concentration (Rockel et al. 2002). In addition to NR, a plasma membrane-bound enzyme of tobacco roots was shown to catalyze the formation of NO from nitrite. This enzyme is believed to reduce the apoplastic nitrite produced by NR and may play a role in nitrate signalling via NO formation (Stöhr et al. 2001).

5.3.3 NO Signalling in Plant Cells

NO signals are typically designated as either cyclic GMP (cGMP)-dependent or -independent, a tribute to the discovery that one of NO's principal targets is guanylate cyclase. In animals, there are two major families of cGMP-producing guanylate cyclases: the transmembrane receptor class, which contains the guanylate cyclase domain within the intracellular portion of the protein, and the soluble class. The latter is activated by NO. NO activates soluble guanylate cyclase either by binding to the heme iron or through S-nitrosyla-tion at critical cysteine residues (McDonald and Murad 1995).

The occurrence of cGMP in plants has been demonstrated by various mass spectrometry techniques (Brown and Newton 1992). cGMP can stimulate the induction of genes encoding chalcone synthase and ferredoxin NADP$^+$ oxi-doreductase and can initiate anthocyanin biosynthesis in soybean (Bowler et al. 1994). It has been shown that various stimuli, such as gibberellic acid treat-ment of barley aleurone, light stimulation of bean cells or NO treatment of spruce needles can cause transient increases in cGMP levels (Brown and New-ton 1992; Pfeiffer et al. 1994; Penson et al. 1996). The effect of NO-releasing compounds on phytochrome-controlled germination of empress tree seeds has been attributed to NO-activation of cGMP production (Giba 1998; Woj-taszek 2000).

Furthermore, administration of NO donors or recombinant mammalian NOS to tobacco plants or suspension cells triggered expression of the defense genes encoding pathogenesis-related (PR)-1 protein and phenylalanine ammonia lyase (PAL; Fig. 5.2; Durner et al. 1998). These genes were also

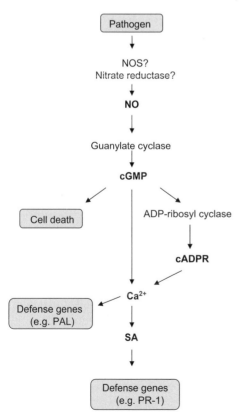

Fig. 5.2. Hypothetical signalling path-way through NO and SA. Shown is the NO-induced expression of defense genes (in this case *PAL* and *PR-1*, two important defense genes in many plants) after pathogen attack. Cur-rently, it is not known how plants pro-duce NO [nitric oxide synthase (*NOS*) or nitrate reductase]. Key enzymes in the signalling cascade are guanylate cyclase and ADP ribosyl cyclase. Important signalling molecules are *NO* (nitric oxide), *cGMP* (cyclic GMP), *cADPR* (cyclic ADP ribose), *Ca*$^{2+}$ and *SA* (salicylic acid). The model suggests SA-dependent as well as SA-independent defense gene induction. For some genes, full activa-tion might require simultaneous pres-ence of both cADPR and cGMP

induced by cGMP and cyclic ADP ribose (cADPR; discussed below), two molecules that can serve as second messengers for NO signalling in mammals. Consistent with cGMP acting as a second messenger in tobacco, NO treatment induced dramatic and transient increases in endogenous cGMP levels similar in magnitude to that reported during NO-induced smooth muscle relaxation in animals (Durner et al. 1998). Moreover, activation of *PAL* gene expression by NO was suppressed by guanylate cyclase inhibitors. It is likely that cGMP is also involved in the induction of *PAL* and chalcone synthase transcripts in NO-treated soybean suspension cells (Delledonne et al. 1998). It should be noted, however, that neither guanylate cyclase nor the phosphodiesterase responsible for cGMP degradation have been cloned from plants, although a cyclic nucleotide phosphodiesterase activity has been described (Reggiani 1997).

Interestingly, cGMP seems to be involved in cell death regulation, too. Specific inhibitors of guanylate cyclase blocked NO-induced cell death in *Arabidopsis* cells, and this inhibition was reversed by the cell-permeable cGMP analogue, 8Br-cGMP, although 8Br-cGMP alone did not induce cell death or potentiate NO-induced cell death. This suggests that cGMP synthesis is required, but not sufficient for NO-induced cell death in *Arabidopsis* (Clarke et al. 2000). Other evidence for the involvement of cGMP in regulation of cell death was obtained after analyzing the *Arabidopsis dnd1* mutant, which was previously isolated as a line that failed to produce the HR in response to avirulent *Pseudomonas syringae* pathogens. DND1 encodes a cyclic nucleotide-gated ion channel that can allow passage of Ca and other cations. This demonstrates that broad-spectrum disease resistance and inhibition of the HR can be activated in plants by disruption of a cGMP-gated ion channel (Clough et al. 2000).

The involvement of cyclic ADP ribose (cADPR) as a second messenger mediating NO effects in plants has been tentatively assigned. In plants, cADPR has been shown to participate in the abscisic acid signalling pathway, probably through the modulation of vacuolar RYR-like proteins(Allen et al. 1995). Moreover, Ca^{2+} fluxes play key roles during the early stages of plant defense responses against pathogens (Scheel 1998). It was found that cADPR induces *PAL* and *PR-1* expression, and this induction is blocked by ruthenium red, an RYR inhibitor. In NO-treated plants, the cADPR antagonist 8-bromo-cADPR suppresses *PR-1* activation, suggesting that NO acts, at least in part, through cADPR-dependent signalling (Klessig et al. 2000). Interestingly, cGMP and cADPR appear to act synergistically to activate the expression of these two genes, just as reported for gene activation in animal cells. In addition, NO activation of defense genes might occur through more than one pathway. For example, NO could directly regulate RYR-like proteins through *S*-nitrosation (Xu et al. 1998).

It should be noted that cADPR is a signalling molecule involved in several cellular functions not related to plant defense reactions. However, recent evidence points to NO/cADPR signalling in ABA-induced stomatal closure. NO synthesis was required for ABA-induced closure and ABA enhanced NO synthesis in guard cells. Exogenous NO induced stomatal closure, and ABA and NO-induced closure required the synthesis and action of cGMP and cADPR (Mata and Lamattina 2001; Neill et al. 2002).

5.3.4 Cross-Talk of NO with Reactive Oxygen Species and SA

A growing body of evidence suggests that ROS appear to play key roles in early and later stages of the plant response. ROS may act as both cellular signals and direct weapons. However, the participation of ROS in induction of host cell death and pathogen killing might be a necessary, but not sufficient, requirement (Dangl 1998). In animals, ROS collaborate with NO to induce apoptosis and to execute invading pathogens (Mayer and Hemmens 1997; Brüne et al. 1998). The reaction of NO· with O_2^- yields peroxynitrite (ONOO−). ONOO− is a powerful oxidant and cytotoxic species implicated in a number of pathophysiological processes. ONOO− reacts to a large range of biomolecules, including proteins, nucleic acids and lipids (Hippeli and Elstner 1996, 1998). Oxidative reaction products in proteins primarily include modifications of cysteine, methionine, tryptophan and tyrosine residues. Irreversible tyrosine nitration by ONOO− may interfere with signal transduction pathways involving tyrosine phosphorylation. Moreover, at physiological pH, ONOO− equilibrates rapidly with pernitrous acid (ONOOH). This latter rearranges to NO_3^- in competition with a reaction that has been thought to involve decomposition to NO_2· and the radical oxidant hydroxyl radical (HO·), the most reactive species known to chemistry (Wink and Mitchell 1998). Interestingly, because of the capacity of NO· to combine to O_2^-, it has been suggested that the production of ONOO− may also represent a protective mechanism against O_2^- toxicity. According to Halliwell et al. ONOO− formation may provide a mechanism of using NO· to dispose of excess O_2^- (or of using O_2^- to dispose of excess NO·) in order to maintain the correct balance between these radicals in vivo (Halliwell et al. 1995).

In plants, it has been recently demonstrated (1) that NO acts synergistically with ROS to increase host cell death of soybean suspension cells and (2) that inhibitors of NOS compromise the hypersensitive resistance response in *Arabidopsis* and tobacco (Delledonne et al. 1998; Huang and Knopp 1998). Strikingly, others have reported a protective role of NO during ROS-mediated cell death (Beligni and Lamattina 1999). The rates of production and dismutation/scavenging of ROS and NO generated during the oxidative burst seems play a crucial role in the modulation and integration of NO/ROS signalling in the HR (Delledonne et al. 2001). In soybean, the efficient induction of hypersensitive cell death requires a balance between ROIs and NO production such that high levels of NO are ineffective in the absence of a correspondingly strong oxidative burst. ONOO− is not an effective inducer of cell death. H_2O_2,

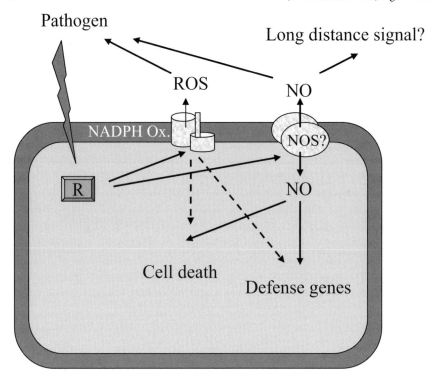

Fig. 5.3. NO burst and hypothetical modes of action of NO in plant defense responses. Invading pathogens are recognized by a host-encoded disease resistance gene product (*R*). This activates a signal transduction cascade, which includes an oxidative burst leading to production of reactive oxygen species (*ROS*). Superoxide (O_2^-) is probably generated by NADPH oxidase. Nitric oxide (*NO*) is produced by a nitric oxide synthase-like enzyme (*NOS*), or by alternative sources. ROS and NO collaborate to execute pathogens via formation of highly toxic peroxynitrite (*ONOO$^-$*). Intracellular NO effects include transcriptional activation of defense genes and induction of hypersensitive host cell death (apoptosis). Intra- or extracellular NO might serve as a long distance, mobile signal for development of systemic resistance

formed by SOD-catalyzed dismutation of O_2^-, functions with NO in the triggering of plant HR (Fig. 5.3). The rates of production and dismutation of O_2^- generated during the pathogen-induced oxidative burst seem to play a crucial role in modulating and integrating the binary NO/H_2O_2 trigger. Delledonne et al. concluded that the functional interactions between NO and ROIs in the plant HR are strikingly different from those observed previously in the vertebrate native immune system (Delledonne et al. 2001). However, these results, together with the discovery of plant homologues of the mammalian NADPH oxidase (Scheel 1998), suggest that the two key enzymes involved in mammalian macrophage and neutrophil action, NADPH oxidase and NOS, are activated during plant-pathogen interactions.

A relationship between NO and SA is also apparent in plants and appears to be multifaceted as well. For example, NO seems to act, at least partially, through a SA-dependent signalling pathway. NO treatment of tobacco leaves induced a significant increase in endogenous SA, which was required for *PR-1* gene induction (Durner et al. 1998). In addition, NO potentiates soybean defense responses by synergizing not only with ROS, but also with SA (or a factor downstream of SA) (Delledonne et al. 1998). Recent results on SAR induction in tobacco have confirmed that NO is required for the full function of SA as an SAR inducer. The activity of NO was fully dependent on the function of SA in the SAR signalling pathway in tobacco (Song and Goodman 2001). Nevertheless, the relations among NO, SA and ROS in the activation of defense genes and/or induction of host cell death is currently unresolved. However, several models suggest a self-amplifying process during which redox signalling through NO and ROS is enhanced by SA (Richberg et al. 1998; Van Camp et al. 1998). This view is supported by the earlier finding that SA treatment leads to enhanced NO production in soybean (Klepper 1991). SA could also mediate and/or potentiate the effects of NO since many NO-regulated enzymes, including aconitase, catalase and several other heme proteins, are likewise regulated by SA (Stamler 1994; Durner et al. 1997).

The above results implicate SA as a co-conspirator with NO, where SA mediates, potentiates and/or facilitates the effects of NO in various processes. However, SA might also counter NO and its effects. Interestingly, one of the most important anti-oxidative (and/or anti-inflammatory) functions of salicylates in mammalian cells is to inhibit the activity and transcription of iNOS (Farivar and Brecher 1996). Furthermore, SA is a potent scavenger of NO and its derivatives (Hermann et al. 1999). Finally, while NO is a potent inhibitor of the respiratory cytochrome c oxidases of both plant and mammalian mitochondria (Millar and Day 1997), plants are well adapted to the toxic effects of NO. In animals, inhibition of cytochrome c oxidase may increase ROS production, another putative link between NO and oxidative stress. In tobacco, SA induces alternative oxidase, which in contrast to other respiration chain oxidases of plant and mammalian origin, is insensitive to NO (Millar and Day 1997; Yamasaki 2000; Yamasaki et al. 2001).

Finally, it should be noted that while NO frequently modulates cell death together with redox messengers such as SA or ROS, it might as well act independently of its usual conspirators. The cGMP-mediated cell death has been discussed above (Clarke et al. 2000). Recently, the participation of the mitochondrial permeability transition pore in NO-induced cell death in citrus has been shown (Saviani et al. 2002), and others reported that changes in the gravitational environment of *Kalanchoe daigremontiana* caused a NO burst, which independently of ROS induced a sequence of cellular events that ultimately led to apoptosis (Pedroso and Durzan 1999; Pedroso et al. 2000).

5.4 Plant Responses to the Environment

5.4.1 Effects of Atmospheric and Soil-borne NO on Plants

So far we have focused on endogenously generated NO as a byproduct of nitrogen assimilation or as a signalling molecule during pathogen attack. However, NO is also a constituent of polluted air. Ozone, oxidative stress and atmospheric pollutants reduce growth of plants in greenhouses and contribute to the decline of trees (and other plants) in urban as well as forest environments (Skelly and Innes 1995; Hufton et al. 1996; Sandermann 1998). While NO and NO_2 rarely cause visible injury, NO pollution can significantly reduce tree life (Wellburn and Wellburn 1996). The NO-dependent inhibition of photosynthesis, transpiration and respiration is diagnostic for selecting spruce trees against decline and potted plants against air pollution in nurseries and greenhouse environments (Wellburn 1990; Takahashi and Yamasaki 2002). The combined stresses of resisting cellular acidification, enhanced levels of nutrients (nitrite and ammonia) and the direct interference of NO with critical metal-containing enzymes and reaction centers (Clark et al. 2000) are thought to be major reasons why nitrogen oxides, especially NO, show a deleterious effect on plant growth and development. In the presence of other air pollutants such as SO_2, NO and NO_2 cause free radical-induced injury similar to that induced by ozone (Wellburn 1990). On the other hand, atmospheric NO might play a role as mediator of plant defense reactions just as reported for reactive oxygen species (Sandermann 2000).

In addition to its role as an atmospheric pollutant, NO may participate in seed germination after forest fires. Many trees release seeds during fire that subsequently germinate in response to components of wood smoke. In some cases (for example the Empress tree, *Paulownia tomentosa*), the germination-inducing component has been clearly identified as NO (Giba 1998). In many plants, germination is dependent on nitrite or hydroxylamines, whose chelation with iron atoms of hemes is an endogenous source of NO (Hendricks and Taylerson 1975). Interestingly, in those experiments breaking of seed dormancy was caused by catalase inhibition, an effect that can be attributed to NO (Hendricks and Taylerson 1975; Clark et al. 2000). Together with the fact that NO stimulates light-inducible responses such as de-etiolation and influences hypocotyl elongation (Beligni and Lamattina 2000), these experiments suggest that NO of atmospheric or soil origin can function as a stimulator of germination and photomorphogenesis.

5.4.2 Abiotic Triggers of SA Production

A variety of environmental stimuli were also reported to act on the production of SA. Ultraviolet (UV)-C light or ozone induces biochemical responses in tobacco, similar to those induced by necrotizing pathogens. Exposure of leaves to UV-C light or ozone leads to a transient increase in SA both in the exposed as well as in the unexposed leaves from exposed plants. The accumulation in SA levels are accompanied by an increase in SA conjugate, in benzoic acid 2-hydroxylase activity as well as in PR1 accumulation. This was paralleled by an increase in SAR to a subsequent challenge with tobacco mosaic virus. UV light, ozone fumigation and tobacco mosaic virus can activate a common signal transduction pathway that leads to SA and PR-protein accumulation and SAR (Yalpani et al. 1994). In *Arabidopsis*, both UV-C and ozone treatment strongly induce the accumulation of SA and SA-conjugate (Nawrath and Métraux 1999). It seems that ethylene and JA signalling pathways play opposing roles in ozone- and superoxide-induced ROS accumulation and spreading lesions. These findings are based on the study of an ozone-sensitive mutant of *Arabidopsis* (*rcd1*). Ethylene perception and signalling are likely to be involved in promoting cell death, while JA signalling might be responsible for the lesion containment (Overmyer et al. 2000). Application of UV was also tested as a means to control storage disease caused by *Botrytis cinerea* in carrot roots (Mercier et al. 2000). UV radiation acted only in the exposed tissue and induced a local accumulation of 6-methoxymellein (6-MM) reaching levels equal to or higher than the ED_{50} (50 % effective dose) for inhibiting *B. cinerea* at the time of challenge. UV treatment also primed the root tissue for a stronger induction of chitinase after challenge with *B. cinerea* (Mercier et al. 2000). UV-C treatment also controls *B. cinerea* in bell pepper by an effect on the resistance on the plant tissue (Mercier et al. 2001). In this cases it is not known if UV-C also induces the levels of endogenous SA that could be involved in the activation of induced resistance.

Light, in particular light perceived by the phytochrome photoreceptors was also found to be required for the development of SAR in *Arabidopsis*. The *phyA* and *phyB* light receptor mutants or the *phyA/phyB* double mutants are strongly impaired in SA-induced *PR-1* and resistance. Light seems to enhance the sensitivity of the tissue to its own SA rather than stimulate SA production and the light pathway connects to the SA pathway downstream of SA accumulation (Genoud et al. 2002). The effect of light appears to be specific for the SA-dependent signalling pathway only.

SA added to the hydroponic growth solution of young maize plants under normal growth conditions provided protection against subsequent low-temperature stress. It was proposed that this effect might result from the induction of antioxidative enzymes that lead to chilling resistance (Janda et al. 1999). In the same material, chilling at 5 °C caused an increase in aminocyclopropane-1-carboxylic acid content; however, this increase was less pronounced in cold-acclimated plants after pretreatment with SA before the cold stress (Szalai et al. 2000). In barley, SA and aspirin were found to induce the accumulation of glycinebetaine, an osmoprotectant produced in response to cold, drought and osmotic stress (Jagendorf and Takabe 2001). Signalling involved in the plant adaptive responses to salinity/drought stresses is known to involve reversible protein phosphorylation/dephosphorylation. In tobacco cells, two MAPKs, identified as SIPKs (SA-induced protein kinase), are activated in response to hyperosmotic stress induced by NaCl. One of these SIPKs is a 40-kD protein that is specific for the hyperosmotic stress and is Ca^{2+}- and ABA-independent (Hoyos and Zhang 2000).

These examples illustrate some intriguing relations between SA and the responses of plants to a number of environmental stresses. Like in the interactions with biotic stresses, SA presumably acts in a signalling cascade leading to the stress response. The future challenge will be to elucidate how these signalling routes are organized in a regulatory network that copes with various stresses that often can arise simultaneously.

5.5 Conclusions

It is becoming increasingly apparent that SA and NO are ubiquitous signals and effectors in plants, just as reported for mammals. SA plays a key role in induction of SAR. However, NO, or one or more derivatives of it, may also play a role in long-distance signalling, perhaps in SAR. Considering the interactions among the pleiotropic effectors NO, SA and ROS that are currently emerging, it is apparent that we are at the very early stages of understanding the involvement of SA and NO in the action of biotic and abiotic stressors in plants.

References

Allan AC, Fluhr R (1997) Two distinct sources of elicited reactive oxygen species in tobacco epidermal cells. Plant Cell 9:1559–1572

Allen GJ, Muir SR, Sanders D (1995) Realease of Ca^{2+} from individual plant vacuoles by both InsP3 and cyclic ADP-ribose. Science 268:735–737

Asai T, Tena G, Plotnikova J, Willmann MR, Chiu WL, Gomez-Gomez L, Boller T, Ausubel F M, Sheen J (2002) MAP kinase signalling cascade in *Arabidopsis* innate immunity. Nature 415:977–983

Audenaert K, De Meyer GB, Hofte MM (2002) Abscisic acid determines basal susceptibility of tomato to Botrytis cinerea and suppresses salicylic acid-dependent signaling mechanisms. Plant Physiol 128:491–501

Baeuerle P, Baltimore D (1996) NF-kB: ten years after. Cell 87:13–20

Baker C, Orlandi E (1995) Active oxygen in plant pathogenesis. Annu Rev Phytopathol 33:299–321

Beers EP, McDowell JM (2001) Regulation and execution of programmed cell death in response to pathogens, stress and developmental cues. Curr Opin Plant Biol 4:561–567

Beligni MV, Lamattina L (1999) Nitric oxide counteracts cytotoxic processes mediated by reactive oxygen species in plant tissues. Planta 208:337–344

Beligni MV, Lamattina L (2000) Nitric oxide stimulates seed germination and de-etiolation, and inhibits hypocotyl elongation, three light-inducible responses in plants. Planta 210:215–221

Bender J, Fink GR (1998) A Myb homologue, ATR1, activates tryptophan gene expression in *Arabidopsis*. Proc Natl Acad Sci USA 95:5655–5660

Berrocal-Lobo M, Molina A, Solano R (2002) Constitutive expression of ETHYLENE-RESPONSE-FACTOR1 in *Arabidopsis* confers resistance to several necrotrophic fungi. Plant J 29:23–32

Bittner-Eddy PD, Beynon JL (2001) The *Arabidopsis* downy mildew resistance gene, RPP13-Nd, functions independently of NDR1 and EDS1 and does not require the accumulation of salicylic acid. Mol Plant-Microbe Interact 14:416–421

Boller T, Gehri A, Mauch F, Vögeli U (1983) Chitinase in bean leaves: induction by ethylene, purification, properties, and possible function. Planta 157:22–31

Bolwell GP, Davies DR, Gerrish C, Auh CK, Murphy TM (1998) Comparative biochemistry of the oxidative burst produced by rose and French bean cells reveals two distinct mechanisms. Plant Physiol 116:1379–1385

Bostock RM, Karban R, Thaler JS, Weyman PD, Gilchrist D (2001) Signal interactions in induced resistance to pathogens and insect herbivores. Eur J Plant Pathol 107:103–111

Bowler C, Neuhaus G, Yamagata H, Chua N-H (1994) Cyclic GMP and calcium mediate phytochrom transduction. Cell 77:73–81

Bowling SA, Guo A, Cao H, Gordon AS, Klessig DF, Dong XN (1994) A mutation in *Arabidopsis* that leads to constitutive expression of systemic acquired-resistance. Plant Cell 6:1845–1857

Bowling SA, Clarke JD, Liu YD, Klessi, DF, Dong XN (1997) The cpr5 mutant of *Arabidopsis* expresses both NPR1-dependent and NPR1-independent resistance. Plant Cell 9:1573–1584

Bredt DS, Snyder SH (1990) Isolation of nitric oxide synthetase, a calmodulin-requiring enzyme. Proc Natl Acad Sci USA 87:682–685

Brisson LF, Tenhaken R, Lamb C (1994) Functions of oxidative cross-linking of cell wall structural proteins in plant disease resistance. Plant Cell 6:1703–1712

Brüne B, von Knethen A, Sandau KB (1998) Nitric oxide and its role in apoptosis. Eur J Pharmacol 351:261–272

Brown EG, Newton RP (1992) Analytical procedures for cyclic nucleotides and their associated enzymes in plant tissue. Phytochem Anal 3:1–13

Brown MH, Paulsen IT, Skurray RA (1999) The multidrug efflux protein NorM is a prototype of a new family of transporters. Mol Microbiol 31:394–395

Cao H, Glazebrook J, Clarke JD, Volko S, Dong XN (1997) The *Arabidopsis* NPR1 gene that controls systemic acquired resistance encodes a novel protein containing ankyrin repeats. Cell 88:57–63

Carbone D, Arnaudi C (1930) L'immunità nelle piante. Monografie dell'Istituto Sieroterapico Milanese, Milano

Cardinale F, Jonak C, Ligterink W, Niehaus K, Boller T, Hirt H (2000) Differential activation of four specific MAPK pathways by distinct elicitors. J Biol Chem 275:36734–36740

Chen ZX, Ricigliano JW, Klessig DF (1993) Purification and characterization of a soluble salicylic acid- binding protein from tobacco. Proc Natl Acad Sci USA 90:9533–9537

Chester KS (1933) The problem of acquired physiological immunity in plants. Quart Rev Biol 8:275–324

Chong J, Pierrel MA, Atanassov R, Werck-Reithhart D, Fritig B, Saindrenan P (2001) Free and conjugated benzoic acid in tobacco plants and cell cultures. induced accumulation upon elicitation of defense responses and role as salicylic acid precursors. Plant Physiol 125:318–328

Ciardi JA, Tieman DM, Jones JB, Klee HJ (2001) Reduced expression of the tomato ethylene receptor gene LeETR4 enhances the hypersensitive response to *Xanthomonas campestris* pv. *vesicatoria*. Mol Plant Microbe Interact 14:487–495

Clark D, Durner J, Navarre DA, Klessig DF (2000) Nitric oxide inhibition of tobacco catalase and ascorbate peroxidase. Mol Plant Microbe Interact 13:1380–1384

Clarke A, Desikan R, Hurst RD, Hancock JT, Neill SJ (2000) NO way back: nitric oxide and programmed cell death in *Arabidopsis thaliana* suspension cultures. Plant J 24:667–677

Clarke JD, Liu YD, Klessig DF, Dong XN (1998) Uncoupling PR gene expression from NPR1 and bacterial resistance: characterization of the dominant *Arabidopsis* cpr6-1 mutant. Plant Cell 10:557–569

Clough SJ, Fengler KA, Yu IC, Lippok B, Smith RK Jr, Bent AF (2000) The *Arabidopsis* dnd1 "defense, no death" gene encodes a mutated cyclic nucleotide-gated ion channel. Proc Natl Acad Sci USA 97:9323–9328

Cohn J, Sessa G, Martin GB (2001) Innate immunity in plants. Curr Opin Immunol 13:55–62

Conrath U, Pieterse C, Mauch-Mani B (2002) Priming inplant-pathogen interactions. Trends Plant Sci 7:210–216

Cooney RV, Harwood PJ, Custer LJ, Franke AA (1994) Light-mediated conversion of nitrogen dioxide to nitric oxide by carotenoids. Environ Health Perspect 102:460–462

Creelman RA, Mullet JE (1997) Biosynthesis and action of jasmonates in plants. Annu Rev Plant Physiol Plant Mol Biol48:355–381

Cueto M, Hernand'z-Perera O, Martin R, Bentura ML, Rodrigo J, Lamas S, Golvano MP (1996) Presence of nitric oxide synthase activity in roots and nodules of Lupinus albus. FEBS Lett 398:159–164

Dangl J (1998) Plants just say NO to pathogens. Nature 394:525–526

Dangl JL, Jones JDG (2001) Plant pathogens and integrated defence responses to infection. Nature 411:826–833

Debeaujon I, Peeters AJM, Leon-Kloosterziel KM, Koornneef M (2001) The TRANSPARENT TESTA12 gene of *Arabidopsis* encodes a multidrug secondary transporter-like protein required for flavonoid sequestration in vacuoles of the seed coat endothelium. Plant Cell 13:853–871

Delaney TP, Uknes S, Vernooij B, Friedrich L, Weymann K, Negretto D, Gaffney T, Gut-Rella M, Kessmann H, Ward E, Ryals J (1994) A central role of salicylic acid in plant disease resistance. Science 266:1247–1249

Delledonne M, Xia Y, Dixon RA, Lamb C (1998) Nitric oxide functions as a signal in plant disease resistance. Nature 394:585–588

Delledonne M, Zeier J, Marocco A, Lamb C (2001) Signal interactions between nitric oxide and reactive oxygen intermediates in the plant hypersensitive disease resistance response. Proc Natl Acad Sci USA 98:13454–13459

Despres C, DeLong C, Glaze S, Liu E, Forbert PR (2000) The *Arabidopsis* NPR1/MIM1 protein interacts with a subgroup of the TGA family of bZIP transcription factors. Plant Cell 12:279–290

Dewdney J, Reuber TL, Wildermuth MC, Devoto A, Cui JP, Stutius LM, Drummond EP, Ausubel FM (2000) Three unique mutants of *Arabidopsis* identify eds loci required for limiting growth of a biotrophic fungal pathogen. Plant J 24:205–218

Diener AC, Gaxiola RA, Fink GR (2001) *Arabidopsis* ALF5, a multidrug efflux transporter gene family member, confers resistance to toxins. Plant Cell 13:1625–1637

Dinesh-Kumar SP, Tham WH, Baker BJ (2000) Structure-function analysis of the tobacco mosaic virus resistance gene N. Proc Natl Acad Sci USA 97:14789–14794

Doke N (1997) The oxidative burst: roles in signal transduction and plant stress. In: Scandalios JG (ed) Oxidative stress and the molecular biology of antioxidant defenses. Cold Spring Harbor Laboratory Press, Plainview, pp 785–813

Du H, Klessig DF (1997) Identification of a soluble, high-affinity salicylic acid- binding protein in tobacco. Plant Physiol 113:1319–1327

Durner J, Klessig DF (1995) Inhibition of ascorbate peroxidase by salicylic acid and 2,6-dichloroisonicotinic acid, two inducers of plant defense responses. Proc Natl Acad Sci USA 92:11312–11316

Durner J, Klessig DF (1996) Salicylic acid is a modulator of tobacco and mammalian catalases. J Biol Chem 271:28492–28501

Durner J, Klessig DF (1999) Nitric oxide as a signal in plants. Curr Opin Plant Biol 2:369–372

Durner J, Shah J, Klessig DF (1997) Salicylic acid and disease resistance in plants. Trends Plant Sci 2:266–274

Durner J, Wendehenne D, Klessig DF (1998) Defense gene induction in tobacco by nitric oxide, cyclic GMP, and cyclic ADP-ribose. Proc Natl Acad Sci USA 95:10328–10333

Durner J, Gow AJ, Stamler JS, Glazebrook J (1999) Ancient origins of nitric oxide signaling in biological systems. Proc Natl Acad Sci USA 96:14206–14207

Ellis C, Turner JG (2001) The *Arabidopsis* mutant cev1 has constitutively active jasmonate and ethylene signal pathways and enhanced resistance to pathogens. Plant Cell 13:1025–1033

Ellis JG, Lawrence GJ, Luck JE, Dodds PN (1999) Identification of regions in alleles of the flax rust resistance gene L that determine differences in gene-for-gene specificity. Plant Cell 11:495–506

Eulgem T, Rushton PJ, Robatzek S, Somssich IE (2000) The WRKY superfamily of plant transcription factors. Trends Plant Sci 5:199–206

Fan J, Frey RS, Rahman A, Malik AB (2001) Role of neutrophil NADPH oxidase in the mechanism of TNFa-induced NF-kB activation and ICAM-1 expression in endothelial cells. J Biol Chem 29:29

Farivar RS, Brecher P (1996) Salicylate is a transcriptional inhibitor of the inducible nitric oxide synthase in cultured cardiac fibroblasts. J Biol Chem 271:31585–31592

Farmer EE, Weber H, Vollenweider S (1998) Fatty acid signaling in *Arabidopsis*. Planta 206:167–174

Felix G, Grosskopf DG, Regenass M, Basse CW, Boller T (1991) Elicitor-induced ethylene biosynthesis in tomato cells – characterization and use as a bioassay for elicitor action. Plant Physiol 97:19–25

Felton GW, Korth KL, Bi JL, Wesley SV, Huhman DV, Mathews MC, Murphy JB, Lamb C, Dixon RA (1999) Inverse relationship between systemic resistance of plants to microorganisms and to insect herbivory. Curr Biol 9:317–320

Feys BJF, Benedetti CE, Penfold CN, Turne JG (1994) *Arabidopsis* mutants selected for resistance to the phytotoxin coronatine are male-sterile, insensitive to methyl jasmonate, and resistant to a bacterial pathogen. Plant Cell 6:751–759

Foissner I, Wendehenne D, Langebartels C, Durner J (2000) In vivo imaging of an elicitor-induced nitric oxide burst in tobacco. Plant J 23:817–824

Frye CA, Tang DZ, Innes RW (2001) Negative regulation of defense responses in plants by a conserved MAPKK kinase. Proc Natl Acad Sci USA 98:373–378

Gäumann E (1946) Pflanzliche Infektionslehre. Birkhäuser, Basel

Gaffney T, Friedrich L, Vernooij B, Negrotto D, Nye G, Uknes S, Ward E, Kessmann H, Ryals J (1993) Requirement of salicylic acid for the induction of systemic acquired resistance. Science 261:754–756

Genoud T, Buchala A, Chua NH, Métraux JP (2002) Phytochrome signaling modulates the SA-perceptive pathway in *Arabidopsis*. Plant J 31:87–95

Giba Z (1998) The effect of NO-releasing compounds on phytochrome-controlled germination of empress tree seeds. Plant Growth Reg 26:175–181

Glazebrook J (2001) Genes controlling expression of defense responses in *Arabidopsis* – 2001 status. Curr Opin Plant Biol 4:301–308

Govrin EM, Levine A (2000) The hypersensitive response facilitates plant infection by the necrotrophic pathogen *Botrytis cinerea*. Curr Biol 10:751–757

Greenberg JT, Guo AL, Klessig DF, Ausubel FM (1994) Programmed cell death in plants: a pathogen-triggered response activated coordinately with multiple defense functions. Cell 77:551–563

Greenberg JT, Silverman FP, Liang H (2000) Uncoupling salicylic acid-dependent cell death and defense- related responses from disease resistance in the *Arabidopsis* mutant acd5. Genetics 156:341–350

Gupta V, Willits MG, Glazebrook J (2000) *Arabidopsis thaliana* EDS4 contributes to salicylic acid (SA)- dependent expression of defense responses: evidence for inhibition of jasmonic acid signaling by SA. Mol Plant Microbe Interact 13:503–511

Halliwell B, Gutteridge JMC (1989) Free radicals in biology and medicine. Clarendon Press, Oxford

Halliwell B, Aeschbach R, Löliger J, Aruoma OI (1995) The characterization of antioxidants. Food Chem Toxic 33:601–617

Hammond-Kosack KE, Jones JDG (1996) Resistance gene-dependent plant defense responses. Plant Cell 8:1773–1791

Heil M, Baldwin IT (2002) Fitness costs of induced resistance: emerging experimental support for a slippery concept. Trends Plant Sci 7:61–67

Hendricks SB, Taylerson RB (1975) Breaking of seed dormancy by catalase inhibition. Proc Natl Acad Sci USA 72:306–309

Hermann M, Kapiotis S, Hofbauer R, Exner M, Seelos C, Held I, Gmeiner B (1999) Salicylate inhibits LDL oxidation initiated by superoxide/nitric oxide radicals. FEBS Lett 45:212–214

Hevel JM, Marletta MA (1994) Nitric oxide synthase assays. Methods Enzymol 233:250–258

Hippeli S, Elstner EF (1996) Mechanisms of oxygen activation during plant stress: biochemical effects of air pollutants. J Plant Physiol 148:249–257

Hippeli S, Elstner EF (1998) Transition metal ion-catalyzed oxygen activation during pathogenic processes. FEBS Lett 443:1–7

Hirt H (1997) Multiple roles of MAP kinases in plant signal transduction. Trends Plant Sci 2:11–15

Hoffman T, Schmidt JS, Zheng XY, Bent AF (1999) Isolation of ethylene-insensitive soybean mutants that are altered in pathogen susceptibility and gene-for-gene disease resistance. Plant Physiol 119:935–949

Hoyos ME, Zhang SQ (2000) Calcium-independent activation of salicylic acid-induced protein kinase and a 40-kilodalton protein kinase by hyperosmotic stress. Plant Physiol 122:1355–1363

Huang J-S, Knopp JA (1998) Involvement of nitric oxide in Ralstonia solanacearum-induced hypersensitive reaction in tobacco. In: Prior P, Elphinstone J, Allen C (eds)

Bacterial wilt disease: molecular and ecological aspects. Springer, Berlin Heidelberg New York, pp 218–224

Hufton CA, Besford RT, Wellburn RA (1996) Effects of NO (+NO2) pollution on growth, nitrate reductase activities and associated protein contents in glasshouse lettuce grown hydroponically in winter with CO2 enrichment. New Phytol 133:495–501

Islam SZ, Honda Y, Arase S (1998) Light-induced resistance of broad bean against *Botrytis cinerea*. J Phytopathol 146:479–485

Jabs T, Dietrich RA, Dangl JL (1996) Initiation of runaway cell death in an *Arabidopsis* mutant by extracellular superoxide. Science 273:1853–1856

Jacobson MD (1996) Reactive oxygen species and programmed cell death. Trends Biochem Sci 21:83–86

Jaffrey SR, Snyder SH (1996) PIN: an associated protein inhibitor of neuronal nitric oxide synthase. Science 274:774–777

Jagendorf AT, Takabe T (2001) Inducers of glycinebetaine synthesis in barley. Plant Physiol 127:1827–1835

Janda T, Szalai G, Tari I, Paldi E (1999) Hydroponic treatment with salicylic acid decreases the effects of chilling injury in maize (*Zea mays* L.) plants. Planta 208:175–180

Jia L, Bonaventura C, Bonaventura J, Stamler JS (1996) *S*-nitrosohaemoglobin: a dynamic activity of blood involved in vascular control. Nature 380:221–226

Katz VA, Thulke OU, Conrath U (1998) A benzothiadiazole primes parsley cells for augmented elicitation of defense responses. Plant Physiol 117:1333–1339

Kauss H, Fauth M, Merten A, Jeblick W (1999) Cucumber hypocotyls respond to cutin monomers via both an inducible and a constitutive H2O2-generating system. Plant Physiol 120:1175–1182

Kinkema M, Fan WH, Dong XN (2000) Nuclear localization of NPR1 is required for activation of PR gene expression. Plant Cell 12:2339–2350

Klepper L (1990) Comparison between NOx evolution mechanisms of wildtype and nr1 mutant soybean leaves. Plant Physiol 93:26–32

Klepper L (1991) NOx evolution by soybean leaves treated with salicylic acid and selected derivatives. Pest Biochem Physiol 39:43–48

Klessig DF, Durner J, Noad R, Navarre DA, Wendehenne D, Kumar D, Zhou JM, Shah J, Zhang S, Kachroo P et al. (2000) Nitric oxide and salicylic acid signaling in plant defense. Proc Natl Acad Sci USA 97:8849–8855

Knoester M, van Loon LC, van den Heuvel J, Hennig J, Bol JF, Linthorst HJM (1998) Ethylene-insensitive tobacco lacks nonhost resistance against soil-borne fungi. Proc Natl Acad Sci USA 95:1933–1937

Kohler A, Schwindling S, Conrath U (2002) Benzothiadiazole-induced priming for potentiated responses to pathogen infection, wounding, and infiltration of water into leaves requires the NPR1/NIM1 gene in *Arabidopsis*. Plant Physiol 128:1046–1056

Kovtun Y, Chiu WL, Tena G, Sheen J (2000) Functional analysis of oxidative stress-activated mitogen- activated protein kinase cascade in plants. Proc Natl Acad Sci USA 97:2940–2945

Lamas S, Perez-Sala D, Moncada S (1998) Nitric oxide: from discovery to the clinic. Trends Pharmacol Sci 19:436–438

Lebel E, Heifetz P, Thorne L, Uknes S, Ryals J, Ward E (1998) Functional analysis of regulatory sequences controlling PR-1 gene expression in *Arabidopsis*. Plant J 16:223–233

Leshem YY (1996) Nitric oxide in biological systems. Plant Growth Regul 18:155–159

Levine A, Tenhaken R, Dixon R, Lamb C (1994) H2O2 from the oxidative burst orchestrates the plant hypersensitive disease resistance response. Cell 79:583–595

Li X, Zhang YL, Clarke JD, Li Y, Dong XN (1999) Identification and cloning of a negative regulator of systemic acquired resistance, SNI1, through a screen for suppressors of npr1-1. Cell 98:329–339

Lieberherr D, Wagner U, Dubuis PH, Métraux JP, Mauch F (2003) Crosstalk of salicylic acid and ethylene signaling is responsible for the rapid induction of glutathione S-transferase *AtGSTF2* and *AtGSTF6* by avirulent *Pseudomonas syringae*. Plant Cell Physiol (in press)

Luck JE, Lawrence GJ, Dodds PN, Shepherd KW, Ellis JG (2000) Regions outside of the leucine-rich repeats of flax rust resistance proteins play a role in specificity determination. Plant Cell 12:1367–1377

Lund ST, Stal, RE, Klee HJ (1998) Ethylene regulates the susceptible response to pathogen infection in tomato. Plant Cell 10:371-382

Maleck K, Levine A, Eulgem T, Morgan A, Schmid J, Lawton KA, Dangl JL, Dietrich RA (2000) The transcriptome of *Arabidopsis thaliana* during systemic acquired resistance. Nature Genet 26:403–410

Mata CG, Lamattina L (2001) Nitric oxide induces stomatal closure and enhances the adaptive plant responses against drought stress. Plant Physiol 126:1196–204

Mathieu C, Moreau S, Frendo P, Puppo A, Davies MJ (1998) Direct detection of radicals in intact soybean nodules: presence of nitric oxide-leghemoglobin complexes. Free Radic Biol Med 24:1242–1249

Mauch F, Mauch-Mani B, Gaille C, Kull B, Haas D, Reimmann C (2001) Manipulation of salicylate content in *Arabidopsis thaliana* by the expression of an engineered bacterial salicylate synthase. Plant J 25:67–77

Mauch-Mani B, Métraux JP (1998) Salicylic acid and systemic acquired resistance to pathogen attack. Ann Bot 82:535–540

Mauch-Mani B, Slusarenko AJ (1996) Production of salicylic acid precursors is a major function of phenylalanine ammonia-lyase in the resistance of *Arabidopsis* to Peronospora parasitica. Plant Cell 8:203–212

Maxwell DP, Nickels R, McIntosh L (2002) Evidence of mitochondrial involvement in the transduction of signals required for the induction of genes associated with pathogen attack and senescence. Plant J 29:269–279

Mayer B, Hemmens B (1997) Biosynthesis and action of nitric oxide in mammalian cells. Trends Biochem Sci 22:477–481

McDonald LJ, Murad F (1995) Nitric oxide and cGMP signaling, Adv Pharmacol 34:263–276

McDowell JM, Cuzick A, Can C, Beynon J, Dangl JL, Holub EB (2000) Downy mildew (*Peronospora parasitica*) resistance genes in *Arabidopsis* vary in functional requirements for NDR1, EDS1, NPR1 and salicylic acid accumulation. Plant J 22:523–529

Mehdy MC, Sharma YK, Sathasivan K, Bays NW (1996) The role of activated oxygen species in plant disease resistance. Physiol Plant 98:365–374

Mercier J, Lindow SE (2001) Field performance of antagonistic bacteria identified in a novel laboratory assay for biological control of fire blight of pear. Biol Control 22:66–71

Mercier J, Roussel D, Charles MT, Arul J (2000) Systemic and local responses associated with UV- and pathogen- induced resistance to Botrytis cinerea in stored carrot. Phytopathology 90:981–986

Mercier J, Baka M, Reddy B, Corcuff R, Arul J (2001) Shortwave ultraviolet irradiation for control of decay caused by *Botrytis cinerea* in bell pepper: induced resistance and germicidal effects. J Am Soc Hortic Sci 126:128–133

Métraux JP (2001) Systemic acquired resistance and salicylic acid: current state of knowledge. Eur J Plant Pathol 107:13–18

Millar AH, Day DA (1997) Alternative solutions to radical problems. Trends Plant Sci 2:289–290

Mölders W, Buchala A, Métraux JP (1996) Transport of salicylic acid in tobacco necrosis virus-infected cucumber plants. Plant Physiol 112:787–792

Murphy AM, Carr JP (2002) Salicylic acid has cell-specific effects on tobacco mosaic virus replication and cell-to-cell movement. Plant Physiol 128:552–563

Murphy AM, Gilliland A, Eng Wong C, West J, Singh DP, Carr JP (2001) Signal transduction in resistance to plant viruses. Eur J Plant Pathol 107:121–128

Nathan C, Shiloh MU (2000) Reactive oxygen and nitrogen intermediates in the relationship between mammalian hosts and microbial pathogens. Proc Natl Acad Sci USA 97:8841–8848

Nathan C, Xie Q-W (1994) Nitric oxide synthases: roles, tolls, and controls. Cell 78:915–918

Navarre DA, Wendehenne D, Durner J, Noad R, Klessig DF (2000) Nitric oxide modulates the activity of two tobacco enzymes aconitase. Plant Physiol 122:573–582

Nawrath C, Métraux JP (1999) Salicylic acid induction-deficient mutants of *Arabidopsis* express PR-2 and PR-5 and accumulate high levels of camalexin after pathogen inoculation. Plant Cell 11:1393–1404

Nawrath C, Heck S, Parinthawong N, Métraux JP (2002) EDS5, an essential component of salicylic acid-dependent signaling for disease resistance in *Arabidopsis*, is a member of the MATE transporter family. Plant Cell 14:275–286

Neill SJ, Desikan R, Clarke A, Hancock JT (2002) Nitric oxide is a novel component of abscisic acid signaling in stomatal guard cells. Plant Physiol 128:13–16

Ninnemann H, Maier J (1996) Indications for the occurrence of nitric oxide synthases in fungi and plants and the involvement in photoconidation of *Neurospora crassa*. Photochem Photobiol 64:393–398

Noritake T, Kawakita K, Doke N (1996) Nitric oxide induces phytoalexin accumulation in potato tuber tissue. Plant Cell Physiol 37:113–116

Norman-Setterblad C, Vidal S, Palva ET (2000) Interacting signal pathways control defense gene expression in *Arabidopsis* in response to cell wall-degrading enzymes from *Erwinia carotovora*. Mol Plant Microbe Interact 13:430–438

Nürnberger T, Scheel D (2001) Signal transmission in the plant immune response. Trends Plant Sci 6:372–379

Nürnberger T, Nennstiel D, Jabs T, Sacks WR, Hahlbrock K, Scheel D (1994) High affinity binding of a fungal oliopeptide elicitor to parsley plasma membranes triggers multiple defense responses. Cell 78:449–460

Nuhse TS, Peck SC, Hirt H, Boller T (2000) Microbial elicitors induce activation and dual phosphorylation of the *Arabidopsis thaliana* MAPK 6. J Biol Chem 275:7521–7526

Overmyer K, Tuominen H, Kettunen R, Betz C, Langebartels C, Sandermann H, Kangasjarvi J (2000) Ozone-sensitive *Arabidopsis* rcd1 mutant reveals opposite roles for ethylene and jasmonate signaling pathways in regulating superoxide-dependent cell death. Plant Cell 12:1849–1862

Pedroso MC, Durzan D (1999) Detection of apoptosis in chloroplasts and nuclei in different gravitational environments. J Gravit Physiol 6:P19–P20

Pedroso MC, Magalhaes JR, Durzan D (2000) A nitric oxide burst precedes apoptosis in angiosperm and gymnosperm callus cells and foliar tissues. J Exp Bot 51:1027–1036

Pennell RI, Lamb C (1997) Programmed cell death in plants. Plant Cell 9:1157–1168

Penninckx I, Thomma B, Buchala A, Métraux JP, Broekaert WF (1998) Concomitant activation of jasmonate and ethylene response pathways is required for induction of a plant defensin gene in *Arabidopsis*. Plant Cell 10:2103–2113

Penson SP, Schuurink RC, Fath A, Gubler F, Jacobsen JV, Jones RL (1996) cGMP is required for gibberellic acid-induced gene expression in barley aleurone. Plant Cell 8:2325–2333

Petersen M, Brodersen P, Naested H, Andreasson E, Lindhart U, Johansen B, Nielsen HB, Lacy M, Austin MJ, Parker JE, Sharma SB, Klessig DF, Martienssen R, Mattsson O, Jensen AB, Mundy J (2000) *Arabidopsis* MAP kinase 4 negatively regulates systemic acquired resistance. Cell 103:1111–1120

Pfeiffer S, Janystin B, Jessner G, Pichorner H, Ebermann R (1994) Gaseous nitric oxide stimulates guanosine-3':5'-cyclic monophosphate (cGMP) formation in spruce needles. Phytochemistry 36:259–262

Pieterse CMJ, van Loon LC (1999) Salicylic acid-independent plant defence pathways. Trends Plant Sci 4:52–58

Pontier D, Gan SS, Amasino RM, Roby D, Lam E (1999) Markers for hypersensitive response and senescence show distinct patterns of expression. Plant Mol Biol 39:1243–1255

Potter S, Uknes S, Lawton K, Winte AM, Chandler D, Dimaio J, Novitzky R, Ward E, Ryals J (1993) Regulation of a Hevein-like gene in *Arabidopsis*. Mol Plant Microbe Interact 6:680–685

Poulos TL, Raman CS, Li H (1998) NO news is good news. Structure 6:255–258

Raskin I (1992) Role of salicylic-acid in plants. Annu Rev Plant Physiol Plant Mol Biol43:439–463

Rate DN, Greenberg JT (2001) The *Arabidopsis* aberrant growth and death mutant shows resistance to *Pseudomonas syringae* and reveals a role for NPR1 in suppressing hypersensitive cell death. Plant J 27:203–211

Reggiani R (1997) Alteration of levels of cyclic nucleotides in response to anaerobiosis in rice sedlings. Plant Cell Physiol 38:740–742

Ribeiro EA, Cunha FQ, Tamashiro WMSC, Martins IS (1999) Growth phase-dependent subcellular localization of nitric oxide synthase in maize cells. FEBS Lett 445:283–286

Richberg MH, Aviv DH, Dangl JL (1998) Dead cells do tell tales. Curr Opin Plant Biol 1:480–485

Rockel P, Strube F, Rockel A, Wildt J, Kaiser WM (2002) Regulation of nitric oxide (NO) production by plant nitrate reductase in vivo and in vitro. J Exp Bot 53:103–10

Roetschi A, Si-Ammour A, Belbahri L, Mauch F, Mauch-Mani B (2001) Characterization of an *Arabidopsis–Phytophthora* pathosystem: resistance requires a functional PAD2 gene and is independent of salicylic acid, ethylene and jasmonic acid signalling. Plant J 28:293–305

Romeis T, Piedras P, Zhang SQ, Klessig DF, Hirt H, Jones JDG (1999) Rapid Avr9- and Cf-9-dependent activation of MAP kinases in tobacco cell cultures and leaves: convergence of resistance gene, elicitor, wound, and salicylate responses. Plant Cell 11:273–287

Ross AF (1966) Systemic effects of local lesion formation. In: Beemster ABR, Dijkstr J (eds) Viruses of plants. North-Holland Publishing, Amsterdam, pp 127–150

Rusterucci C, Montillet JL, Agnel JP, Battesti C, Alonso B, Knoll A, Bessoule JJ, Etienne P, Suty L, Blein JP, Triantaphylides C (1999) Involvement of lipoxygenase-dependent production of fatty acid hydroperoxides in the development of the hypersensitive cell death induced by cryptogein on tobacco leaves. J Biol Chem 274:36446–36455

Ryals JA, Neuenschwander UH, Willits MG, Molina A, Steiner H-Y, Hunt MD (1996) Systemic acquired resistance. Plant Cell 8:1809–1819

Ryals J, Weymann K, Lawton K, Friedrich L, Ellis D, Steiner HY, Johnson J, Delaney TP, Jesse T, Vos P, Uknes S (1997) The *Arabidopsis* NIM1 protein shows homology to the mammalian transcription factor inhibitor I kappa B. Plant Cell 9:425–439

Sandermann H (1998) Ozone: an air pollutant acting as a plant-signaling molecule. Naturwissenschaften 85:369–375

Sandermann H (2000) Active oxygen species as mediators of plant immunity: three case studies. Biol Chem 381:649–653

Sandermann H, Ernst D, Heller W, Langebartels C (1998) Ozone: an abiotic elicitor of plant defense reactions. Trends Plant Sci 3:47–50

Saviani EE, Orsi CH, Oliveira JF, Pinto-Maglio CA, Salgado I (2002) Participation of the mitochondrial permeability transition pore in nitric oxide-induced plant cell death. FEBS Lett 510:136–140

Schaller A, Ryan CA (1996) Systemin–a polypeptide defense signal in plants. Bioessays 18:27–33

Scheel D (1998) Resistance response physiology and signal transduction. Curr Opin Plant Biol 1:305–310

Schenk PM, Kazan K, Wilson I, Anderson JP, Richmond T, Somerville SC, Manners JM (2000) Coordinated plant defense responses in Arabidopsis revealed by microarray analysis. Proc Natl Acad Sci USA 97:11655–11660

Seo HS, Song JT, Cheong JJ, Lee YH, Lee YW, Hwang I, Lee JS, Choi YD (2001) Jasmonic acid carboxyl methyltransferase: a key enzyme for jasmonate-regulated plant responses. Proc Natl Acad Sci USA 98:4788–4793

Serino L, Reimmann C, Baur H, Beyeler M, Visca P, Haas D (1995) Structural genes for salicylate biosynthesis from chorismate in Pseudomonas aeruginosa. Mol Gen Genet 249:217–228

Seskar M, Shulaev V, Raskin I (1998) Endogenous methyl salicylate in pathogen-inoculated tobacco plants. Plant Physiol 116:387–392

Shah J, Kachroo P, Klessig DF (1999) The Arabidopsis ssi1 mutation restores pathogenesis-related gene expression in npr1 plants and renders defensin gene expression salicylic acid dependent. Plant Cell 11:191–206

Shirasu K, Schulze-Lefert P (2000) Regulators of cell death in disease resistance. Plant Mol Biol 44:371–385

Shirasu K, Nakajima H, Rajasekhar VK, Dixon RA, Lamb C (1997) Salicylic acid potentiates an agonist-dependent gain control that amplifies pathogen signals in the activation of defense mechanisms. Plant Cell 9:261–270

Shulaev V, Leon J, Raskin I (1995) Is salicylic acid a translocated signal of systemic acquired resistance in tobacco? Plant Cell 7:1691–1701

Shulaev V, Silverman P, Raskin I (1997) Airborne signalling by methyl salicylate in plant pathogen resistance. Nature 386:738–738

Skelly JM, Innes JL (1995) Waldsterben in forests of central Europe and Esatern North America: fantasy or reality? Plant Disease 78:1021–1032

Song F, Goodman RM (2001) Activity of nitric oxide is dependent on, but is partially required for function of, salicylic acid in the signaling pathway in tobacco systemic acquired resistance. Mol Plant Microbe Interact 14:1458–1462

Stöhr C, Strube F, Marx G, Ullrich WR, Rockel P (2001) A plasma membrane-bound enzyme of tobacco roots catalyses the formation of nitric oxide from nitrite. Planta 212:835–841

Stamler JS (1994) Redox signaling: nitrosylation and related target interactions of nitric oxide. Cell 78:931–936

Stamler JS, Singel DL, Loscalzo J (1992) Biochemistry of nitric oxide and its redox-activated forms. Science 258:1898–1902

Staswick PE, Yuen GY, Lehman CC (1998) Jasmonate signaling mutants of Arabidopsis are susceptible to the soil fungus Pythium irregulare. Plant J 15:747–754

Sticher L, Mauch-Mani B, Métraux JP (1997) Systemic acquired resistance. Annu Rev Phytopathol 35:235–270

Stintzi A, Weber H, Reymond P, Browse J, Farmer EE (2001) Plant defense in the absence of jasmonic acid: the role of cyclopentenones. Proc Natl Acad Sci USA 98:12837–12842

Stone JM, Heard JE, Asai T, Ausubel FM (2000) Simulation of fungal-mediated cell death by fumonisin B1 and selection of fumonisin B1-resistant (fbr) Arabidopsis mutants. Plant Cell 12:1811–1822

Summermatter K, Sticher L, Métraux JP (1995) Systemic responses in Arabidopsis thaliana infected and challenged with Pseudomonas syringae pv. syringae. Plant Physiol 108:1379–1385

Szalai G, Tari I, Janda T, Pestenacz A, Paldi E (2000) Effects of cold acclimation and salicylic acid on changes in ACC and MACC contents in maize during chilling. Biol Plant 43:637–640

Takahashi S, Yamasaki H (2002) Reversible inhibition of photophosphorylation in chloroplasts by nitric oxide. FEBS Lett 512:145–148

Tenhaken R, Anstatt C, Ludwig A, Seehaus K (2001) WY-14,643 and other agonists of the peroxisome proliferator-activated receptor reveal a new mode of action for salicylic acid in soybean disease resistance. Planta 212:888–895

Thaler JS, Fidantsef AL, Duffey SS, Bostock RM (1999) Trade-offs in plant defense against pathogens and herbivores: a field demonstration of chemical elicitors of induced resistance. J Chem Ecol 25:1597–1609

Thomma B, Eggermont K, Penninckx I, Mauch-Mani B, Vogelsang R, Cammue BPA, Broekaert WF (1998) Separate jasmonate-dependent and salicylate-dependent defense-response pathways in *Arabidopsis* are essential for resistance to distinct microbial pathogens. Proc Natl Acad Sci USA 95:15107–15111

Thomma B, Penninckx I, Broekaert WF, Cammue BPA (2001) The complexity of disease signaling in *Arabidopsis*. Curr Opin Immunol 13:63–68

Thulke O, Conrath U (1998) Salicylic acid has a dual role in the activation of defence-related genes in parsley. Plant J 14:35–42

Torres M, Onouchi H, Hamada S, Machida C, Hammond-Kosack K, Jones J (1998) Six *Arabidopsis thaliana* homologues of the human respiratory burst oxidase (gp91phox). Plant J 14:365–370

Torres MA, Dangl JL, Jones JD (2002) *Arabidopsis* gp91phox homologues AtrbohD and AtrbohF are required for accumulation of reactive oxygen intermediates in the plant defense response. Proc Natl Acad Sci USA 99:517–522

Van Breusegem F, Vranov E, Dat JF, Inzé D (2001) The role of active oxygen species in plant signal transduction. Plant Sci 161:405–414

Van Camp W, Van Montagu M, Inzé D (1998) H2O2 and NO: redox signals in disease resistance. Trends Plant Sci 3:330–334

Van Loon LC, Van Strien EA (1999) The families of pathogenesis-related proteins, their activities, and comparative analysis of PR-1 type proteins. Physiol Mol Plant Pathol 55:85–97

Vijayan P, Shockey J, Levesque CA, Cook RJ, Browse J (1998) A role for jasmonate in pathogen defense of *Arabidopsis*. Proc Natl Acad Sci USA 95:7209–7214

Waterhouse PM, Wang MB, Lough T (2001) Gene silencing as an adaptive defence against viruses. Nature 411:834–842

Weber H (2002) Fatty acid-derived signals in plants. Trends Plant Sci 7:217–224

Wellburn AR (1990) Why are atmospheric oxides of nitrogen usually phytotoxic and not alternative fertilizers? New Phytol 115:395–429

Wellburn AR, Wellburn FAM (1996) Gaseous pollutants and plant defence mechanisms. Biochem Soc Trans 24:461–464

Wendehenne D, Binet M-N, Blein J-P, Ricci P, Pugin A (1995) Evidence for specific, high - affinity binding sites for a proteinaceous elicitor in tobacco plasma membranes. FEBS Lett 374:203–207

Wendehenne D, Pugin A, Klessig DF, Durner J (2001) Nitric oxide: comparative synthesis and signaling in animal and plant cells. Trends Plant Sci 6:177–183

Weymann K, Hunt M, Uknes S, Neuenschwander U, Lawton K, Steiner HY, Ryals J (1995) Suppression and restoration of lesion formation in *Arabidopsis* lsd mutants. Plant Cell 7:2013–2022

White R (1979) Acetyl salicylic acid (aspirin) induces resistance to tobacco mosaic virus in tobacco. Virology 99:410–412

Whitham S, Dinesh-Kumar SP, Choi D, Hehl R, Corr C, Baker B (1994) The product of the tobacco mosaic virus resistance gene N: similarity to toll and the interleukin-1 receptor. Cell 78:1101–1115

Wildermuth MC, Dewdney J, Wu G, Ausubel FM (2001) Isochorismate synthase is required to synthesize salicylic acid for plant defence. Nature 414:562–565

Wildt J, Kley D, Rockel A, Rockel P, Segschneider HJ (1997) Emission of NO from higher plant species. J Geophys Res 102:5919–5927

Wink DA, Mitchell JB (1998) Chemical biology of nitric oxide: insights into regulatory, cytotoxic, and cytoprotective mechanisms of nitric oxide. Free Rad Biol Med 25:434–456

Wojtaszek P (2000) Nitric oxide in plants. To NO or not to NO. Phytochemistry 54:1–4

Wong CE, Carson RAJ, Carr JP (2002) Chemically induced virus resistance in *Arabidopsis thaliana* is independent of pathogenesis-related protein expression and the NPR1 gene. Mol Plant Microbe Interact 15:75–81

Wrzaczek M, Hirt H (2001) Plant MAP kinase pathways: how many and what for? Biol Cell 93:81–87

Xie ZX, Chen ZX (2000) Harpin-induced hypersensitive cell death is associated with altered mitochondrial functions in tobacco cells, Molecular Plant Microbe Interact 13:183–190

Xing T, Higgins VJ, Blumwald E (1997) Race-specific elicitors of *Cladosporium fulvum* promote translocation of cytosolic components of NADPH oxidase to the plasma membrane of tomato cells. Plant Cell 9:249–259

Xu L, Eu JP, Meissner G, Stamler JS (1998) Activation of the cardiac calcium release channel (ryanodine receptor) by poly-*S*-nitrosylation. Science 279:234–237

Yalpani N, Leon J, Lawton MA, Raskin I (1993) Pathway of salicylic-acid biosynthesis in healthy and virus-inoculated tobacco. Plant Physiol 103:315–321

Yalpani N, Enyedi AJ, Leon J, Raskin I (1994) Ultraviolet-light and ozone stimulate accumulation of salicylic-acid, pathogenesis-related proteins and virus- resistance in tobacco. Planta 193:372–376

Yamasaki H (2000) Nitrite-dependent nitric oxide production pathway: implications for involvement of active nitrogen species in photoinhibition in vivo. Philos Trans R Soc Lond B Biol Sci 355:1477–1488

Yamasaki H, Shimoji H, Ohshiro Y, Sakihama Y (2001) Inhibitory effects of nitric oxide on oxidative phosphorylation in plant mitochondria. Nitric Oxide 5:261–270

Yang KY, Liu YD, Zhang SQ (2001) Activation of a mitogen-activated protein kinase pathway is involved in disease resistance in tobacco. Proc Natl Acad Sci USA 98:741–746

Yang YO, Klessig DF (1996) Isolation and characterization of a tobacco mosaic virus-inducible myb oncogene homolog from tobacco. Proc Natl Acad Sci USA 93:14972–14977

Yoshioka K, Kachroo P, Tsui F, Sharma SB, Shah J, Klessig DF (2001) Environmentally sensitive, SA-dependent defense responses in the cpr22 mutant of *Arabidopsis*. Plant J 26:447–459

Yu IC, Parker J, Bent AF (1998) Gene-for-gene disease resistance without the hypersensitive response in *Arabidopsis* dnd1 mutant. Proc Natl Acad Sci USA 95:7819–7824

Yu IC, Fengler KA, Clough SJ, Bent AF (2000) Identification of arabidopsis mutants exhibiting an altered hypersensitive response in gene-for-gene disease resistance. Mol Plant Microbe Interact 13:277–286

Zhang SQ, Klessig DF (1997) Salicylic acid activates a 48-kD MAP kinase in tobacco. Plant Cell 9:809–824

Zhang S, Klessig DF (2001) MAPK cascades in plant defense signaling. Trends Plant Sci 6:520–527

Zhang S, Liu Y (2001) Activation of salicylic acid-induced protein kinase, a mitogen-activated protein kinase, induces multiple defense responses in tobacco. Plant Cell 13:1877–89

Zhang YL, Fan WH, Kinkema M, Li X, Dong XN (1999) Interaction of NPR1 with basic leucine zipper protein transcription factors that bind sequences required for salicylic acid induction of the PR-1 gene. Proc Natl Acad Sci USA 96:6523–6528

Zhou JM, Trifa Y, Silva H, Pontier D, Lam E, Shah J, Klessig DF (2000) NPR1 differentially interacts with members of the TGA/OBF family of transcription factors that bind an element of the PR-1 gene required for induction by salicylic acid. Mol Plant Microbe Interact 13:191–202

Zimmerli L, Jakab C, Métraux JP, Mauch-Mani B (2000) Potentiation of pathogen-specific defense mechanisms in *Arabidopsis* by b-aminobutyric acid. Proc Natl Acad Sci USA 97:12920–12925

6 *cis* Elements and Transcription Factors Regulating Gene Promoters in Response to Environmental Stress

DIETER ERNST and MARK AARTS

6.1 Transcriptional Regulation

Plants alter their patterns of gene expression in response to environmental stress, and a reprogramming of plant metabolism upon abiotic/biotic stimuli involves changes in plant gene expression. In other words, the expression of many genes involved in primary and secondary metabolism is induced by various ecotoxicological stressors. Similarly, pathogenesis-related genes are induced by a variety of stressors. These changes are the result of a modification in the rate of transcription for specific genes. An eukaroytic gene is divided into several sections. There is a transcribed region, existing of the coding sequence, a 3′ and 5′ untranslated region of mRNA, a promoter region and an upstream regulatory region (Ferl and Paul 2000). The promoter of a single gene can be activated by different external stimuli, but promoters of different genes can also be activated by the same stimuli. Thus, the organization of different promoter sections and the architecture of promoters contribute to the complex gene regulation upon external stressors. The promoter contains several elements important for the transcription of the protein coding DNA sequence. These regulatory elements are called *cis* elements. In general these elements are between 4–20 bp long. Not only their sequence is important for controlling transcription, but also their position relative to the transcription start and relative to other *cis* elements. The basic *cis* elements of eukaryotic genes are the TATA and CAAT or CCAAT boxes. The TATA box is generally found around 30 bp upstream of the transcription start. Together with the CAAT or CCAAT box, located around 70–80 bp upstream of the transcription start, it is involved in directing the RNA polymerase to the initiation of transcription. In addition to these basic elements, inducible genes contain several other *cis* elements, which contribute to the final gene regulation upon various abiotic and biotic stimuli, developmental and environmental signal transduction and also tissue specificity (Fig. 6.1) (Hughes 1996; Bray et al. 2000). Analysis of distinct promoter fragments fused to a reporter gene allows the identification of single or

Ecological Studies, Vol. 170
H. Sandermann (Ed.)
Molecular Ecotoxicology of Plants
© Springer-Verlag Berlin Heidelberg 2004

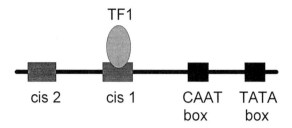

Fig. 6.1. Promoter organization. The most highly conserved *cis* element is the TATA box. Several individual *cis* elements form the basis of a promoter. Transcription factors (*TFs*) interact with these elements to facilitate transcription

multiple promoter sequence elements that affect the level of gene expression. Substitution mutation analysis of such sequence elements finally results in the identification of DNA motifs for individual *cis* elements. Technical terms and methods for the identification of *cis/trans* elements are given in Tables 6.1 and 6.2. Well-known and early characterized *cis* elements are the heat-shock elements (Schöffl et al. 1984), the light-regulated G box and GT-1-binding site (Giuliano et al. 1988; Green et al. 1988) and ABA-responsive sequences (Yamaguchi-Shinozaki et al. 1989; Guiltinan et al. 1990). An increased knowledge of promoters and *cis* elements will become available by the analysis of co-expressed genes on microarrays (Maleck et al. 2000; Seki et al. 2001; Chen et al. 2002). Cluster analysis of co-activated genes upon environmental stress will increase the number of distinct *cis* elements.

Promoter *cis* elements function in concert with *trans*-acting DNA-binding proteins. These transcription factors (*trans* factors) bind to specific *cis* elements, which creates an efficient transcription initiation complex. Thus, the interaction between *cis* elements and *trans* factors is a central step in the regulation of plant gene expression (Fig. 6.1). *Trans* factors typically have at least two domains, one for the recognition and binding to the *cis* element, and the other for organizing additional proteins involved in activating transcription (Ferl and Paul 2000). Transcription factors can be classified according to the properties of their DNA-binding domain. Four major categories, the helix-loop-helix motif, the zinc fingers, the basic leucine zippers and the high mobility group motifs, are known. A higher level of promoter organization exists for promoter modules. Promoter modules are composed of two or more transcription factor-binding sites in a defined distance range, resulting in specific gene activation or repression (Fig. 6.2; Werner 2002). The highest level of organization is the promoter model. Co-regulated genes with functionally related promoters often show defined core transcription factor-binding sites, which are conserved in orientation and distance (Ihmels et al. 2002; Werner 2002).

Genome research of higher plants, such as *Arabidopsis*, rice, barley, soybean and even trees, has produced a large amount of nucleotide sequence information, and there is more to come. Therefore, strategies for the *in silico* identification of promoters are highly desired and in development (e.g., Jensen and

Table 6.1. *cis* elements in stress response in plants. A summary of all motifs and their functions as described in this chapter

Stress	Name	Motif	Reference
UV-B radiation	G box	CACGT	Schulze-Lefert et al. (1989); Menkens et al. (1995)
	GT element	GR(T/A)AA(T/A)	Zhou (1999)
	MRE	ACCTA	Hartmann et al. (1998)
Heat-shock	HSE	nGAAnnTTCnnGAAn	Nover (1987); Almoguera et al. (1998)
Drought and cold	DRE	TATACCGACAT	Baker (1994); Yamaguchi-Shinozaki and Shinozaki (1994)
	CRT	CCGAC	
		CCGAAA	Dunn et al. (1998)
	LTRE	AAGAAGATGC	Brown et al. (2001)
	ABRE	TACGTGGC	Busk and Pagès (1998)
	ABRE	GGACACGTGGC	Busk and Pagès (1998)
	MYC	CACATG	Abe et al. (1997)
	MYB	TGGTTAG	Abe et al. (1997)
Ozone		150 bp promoter region	Schubert et al. (1997)
	W box	(T)TGACT	Ernst et al. (1999)
Xenobiotic	ocs	TGACG(T/C)AAG(CGA/GAC)T(G/T)ACG(TAA/CCC)	Chen and Singh (1999)
	XRE	TnGCGTG	Kernodle and Scandalios (2001)
Heavy metal ions	MRE	TGCRCNC	Zhang et al. (2001a) (animals)
		ACCYYNAAGGT	Lyons et al. (2000) (yeast)
	IDRS	CACGAGNCCNCCAC	Petit et al. (2001)
Elicitor	W box	TGACC/T	Eulgem et al. (1999)
	HSRE	TAAAATTCTTTG	Pontier et al. (2001)
Auxin	AuxRE	TGTCTC	Ulmasov et al. (1997)
Ethylene	GCC box	TAAGAGCCGCC	Ohme-Tagaki and Shinshi (1995)
Cytokinin		1.6 kb promoter	D'Agostino et al. (2000)
Salicylic acid	TCA box	TCATCTTCTT	Goldsbrough et al. (1993)
	W box	TGAC	Yang et al. (1999)
		GGGACTTTTCC	Lebel et al. (1998)
	as-1 box	TGACG(T/C)AAG(C/G(G/A)(A/C)T(G/T)ACG(T/C)(A/C)(A/C)	Zhang and Singh (1994)
Jasmonate	G box	TCACGTGG	Kim et al. (1992)
	C-rich region	CCCTATAGGG	Curtis et al. (1997)
	JASE1	CGTCAATGAA	He and Gan (2001)
	JASE2	CATACGTCGTCAA	He and Gan (2001)

Table 6.2. Methods for the identification of *cis* elements and *trans* factors. (Ausubel et al. 2001; Sambrook and Russell 2001)

Method	Objective
cDNA microarray cluster analysis	*In silico* identification of putative *cis* element motifs
5'-promoter deletion analysis	Crude identification of responsive *cis* element containing region
bp substitution in a motif	Identification of a sequence specific *cis* element
Gel mobility shift assay	Interaction of *cis/trans* elements (responsive DNA region)
DNAse footprinting	Interaction of *cis/trans* elements (responsive DNA region)
Yeast one-hybrid system	Isolation of genes encoding *trans* factors
Yeast two-hybrid system	Isolation of *trans* factor interacting proteins

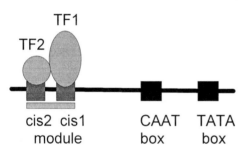

Fig. 6.2. Promoter module. Promoter modules are composed of two or more TF sites, i.e., a combination of *cis* elements in a defined distance range. (Modified from Werner 2002)

Knudsen 2000), especially focused on detection of *cis* elements, including the corresponding *trans* factors. For managing these huge data and information sets, the construction of adequate databases is necessary. The molecular biology database collection, which also includes a *cis*-acting element and *trans* factor database, is regularly updated (Baxevanis 2001; www.nar.oupjournals.org). The eukaryotic promoter database (EPD) is an annotated nonredundant collection of eukaryotic POL II promoters for which the transcription start site has been determined experimentally (Périer et al. 2000; www.epd.isb-sib.ch). The plant *cis*-acting regulatory DNA elements database (PLACE) includes nucleotide sequence motifs found in plant *cis*-acting elements (Higo et al. 1999; www.dna.affrc.go.jp/htdocs/PLACE). PlantCARE is a database of plant *cis*-acting regulatory elements, enhancers and repressors (Rombauts et al. 1999; http://sphinx.rug.ac.be:8080/PlantCARE/index.htm). Four hundred seventeen different names of transcription sites, describing more than 159 plant promoters, are given in PlantCARE. Genomatix (www.genomatix.de) offers MatInspector professional, a tool that utilizes a library of matrix descriptions for transcription factor binding sites to locate matches in sequences of unlimited length (Quandt et al. 1995). The transcription factor database (TRANSFAC) contains information on transcription factors and their origins, functional properties and sequence-specific binding activities (Hehl and Wingender 2001; www.gene-regulation.de).

This chapter will describe important *cis*-acting regulatory elements known to be involved in abiotic environmental stress, as well as elicitor- and hormone-responsive elements. In addition, known *trans* factors interacting with specific *cis* elements are referred to.

6.2 UV-B-Responsive Regulatory Sequences

Stratospheric ozone protects life on earth from deleterious effects of ultraviolet-B (UV-B; 280–320 nm) radiation. In their natural habitat plants are exposed to UV-B radiation, which leads to a wide range of plants responses. The portion of incident UV-B reaching the mesophyll differs considerably between plant species, and significant differences in the epidermal transmittance and in concentrations of UV-B screening compounds between evergreen and deciduous plants exist (Day 1993; Schnitzler et al. 1996). The molecular biology of plants exposed to UV-B has been reviewed by Jordan (1996), and molecular gene markers for UV-B stress in plants have been reported (Brosché et al. 1999). UV-B-induced gene expression focused mainly on phenylpropanoid metabolism, pathogenesis-related proteins, antioxidant enzymes and genes involved in protein turnover (reviewed by Langebartels et al. 2002). Promoter elements involved in regulation of UV-containing white light are well known. The conserved element consists of a CACGTG core sequence and has been studied extensively in the chalcone synthase promoter (Schulze-Lefert et al. 1989; Hartmann et al. 1998; Shimizu et al. 1999). This *cis* element, called G box, is an ubiquitous regulatory element in plants (Menkens et al. 1995), and *trans* factors binding to the G box belong to the bZIP superfamily (Foster et al. 1994; Jakoby and Weisshaaar 2002). G box elements are essential functional components of many stress-responsive promoters and deletion or mutation reduces the overall promoter activity. In addition to the G box, GT elements have been reported to be involved in UV-light-induced gene expression by the binding of GT factors (Ouwerkerk et al. 1999). GT elements show a high degeneracy with a core sequence of G-Pu-(T/A)-A-A-(T/A) (Zhou 1999). GT elements are also found in many promoters that are not regulated by light, indicating that this *cis* element is not just a UV light-specific regulator (Zhou 1999). GT-1 box and G box motifs are also found in photolyase genes (Sakamoto et al. 1998), indicating that an UV-induced accumulation of photolyase transcripts is regulated by transcription factors that interact with these motifs. Another *cis* element involved in UV-signalling is the *MYB* recognition element (*MRE*) with a core sequence of ACCTA (Hartmann et al. 1998). The large family of MYB proteins are involved in the regulation of the phenylpropanoid pathway, and MYB proteins positively regulate the expression of several key phenylpropanoid genes (Weisshaar and Jenkins 1998; Hemm et al. 2001). However, in *Arabidopsis thaliana* the *MYB4* gene is downregulated by UV-B radiation (Jin et al. 2000). MYB4 is a negative regulator of cinnamate 4-hydroxylase, thus increasing the formation of UV-protective sinapate esters. Comparative analysis of the *CHS* promoters from several cruciferous plant species identified conserved UV-responsive regulatory elements, and functional analysis of the *CHS* promoter fragments revealed no major differences in UV light response (Koch et al. 2001). These data indicate, also on a phylogenetic basis, the functional relevance of conserved regulatory promoter motifs.

6.3 Characterization of Heat-Shock Factors

Plants, like many other organisms, in general respond to increases in temperature by the production of Heat Shock Proteins (HSPs). Expression of these HSPs is mainly due to the controlled action of Heat Shock Factors (HSFs), families of transcription factors acting on the promoters of HSP genes. Active HSFs recognize Heat Shock Elements (HSEs) found in promoters of HSP genes and bind to them. The basic structure of HSEs found in many eukaryotes is 5'-AGAAnnTTCT-3' (Nover 1987). In plants, the effective HSE consensus sequence is 5'-nGAAnnTTCnnGAAn-3', although variants containing single or double alternating HSE core consensus sequences (5'-nGAAn-3') or with sequences that are similar but not identical to the consensus core have also been found to be effective (e.g., Almoguera et al. 1998). HSEs are often found in close proximity to the TATA-box element, which may have a functional explanation, considering that HSF1 from *Arabidopsis* can interact with TBP1 (TATA Box Binding Protein1) for rapid activation of HSP transcription (Reindl and Schöffl 1998).

   Contrary to many other eukaryotes, plants contain many genes encoding HSFs. Since the completion of the *Arabidopsis* genome sequence, there are now three classes of plant HSFs known (A, B and C), divided into 14 groups. In *Arabidopsis* currently 21 HSF genes are recognized (Nover et al. 2001), but the NCBI-GenBank database (http://www.ncbi.nlm.nih.gov/entrez/) also contains HSF entries from tomato, tobacco, soybean, tepary bean, alfalfa, pea, rice and corn. The general structure of all HSFs is very similar. Primary amino acid sequences of *Arabidopsis* HSFs vary between 244–495 aa. A DNA-binding domain is found at the N-terminus, followed by a hydrophobic domain involved in protein oligomerization. The C-terminal half contains one or two nuclear localization signal (NLS) motifs. Only in the A-class HSFs, the C-terminus contains an activation domain often characterized by one or two activator or AHA motifs. Some HSFs contain a nuclear export signal (NES) at the C-terminal end. The three HSF classes are clearly distinguished based on their oligomerization signal sequence. This hydrophobic domain consists of an HR-A and an HR-B region. In the A , B and C classes, the two regions are separated by 27, 6 and 13 amino acids, respectively (Nover et al. 2001). Another difference is found in their mode action in DNA binding. A-class HSFs bind as trimers, B-class HSFs probably as dimers (Nover et al. 2001).

Clear differences between HSF genes are found in their expression pattern and localization of the protein. For instance, of the four tomato HSFs studied so far, HSFA1 is constitutively expressed in the cytoplasm (Scharf et al. 1998), HSFA2 is heat stress-induced (Scharf et al. 1998), HSFA3 is constitutively expressed at low levels in the cytoplasm (Bharti et al. 2000) and HSFB1 is transiently expressed in the nucleus upon heat stress (Nover et al. 2001). So far only HSFA proteins appear to be involved in heat-induced gene expression. The function of B- and C-class HSFs is less clear. The *Arabidopsis* B-class AtHSFB1–4 is clearly not able to activate transcription, suggesting that some B-class HSFs may be involved in repression or downregulation of the heat shock response (Czarnecka-Verner et al. 2000). In contrast, preliminary data for tomato HSFB1 suggest that low level co-expression with HSFA1 can result in strong synergistic effects in reporter gene activity (Nover et al. 2001). The absence of AHA motifs in the C-class HSF suggests that this class is also not involved in transcriptional activation (Nover et al. 2001).

6.4 Drought- and Cold-Responsive Gene Regulation

Drought, cold and also high salt often induce the same genes. Some of these genes are also induced by exogenous application of ABA, but others are clearly not. It is the current understanding that both ABA-dependent and ABA-independent pathways are operative (for recent reviews see Thomashow 1999; Shinozaki and Yamaguchi-Shinozaki 2000; Zhu 2001).

Cold and/or drought response elements were already identified in 1994, when Baker et al. (1994) and Yamaguchi-Shinozaki and Shinozaki (1994), respectively, described that many of the cold-induced genes contain one or more CRT (C-repeat)/DRE (dehydration-responsive element) sequences in their promoter region. The full DRE sequence is TACCGACAT, of which the CCGAC core is the designed CRT-sequence. CRT/DRE sequences are recognized by DRE-binding proteins or DREBs, of which five homologues have been reported for *Arabidopsis*, DREB1A, 1B, 1C and DREB2A and 2B (Liu et al. 1998). Of these, DREB1B is identical to CBF1 (CRT-binding factor 1), a transcription factor previously identified to activate CRT/DRE containing promoters (Stockinger et al. 1997). Overexpression of DREB1A (CBF3) also induced expression of a reporter gene driven by a promoter containing a DRE

(Liu et al. 1998). In cDNA-microarray studies for cold- and drought-induced genes (Seki et al. 2001), 12 target genes of DREB1A were identified, 11 of which contained the CCGAC C-repeat. The absence of this sequence in the 2,000 bp upstream region of the FL5–3M24 cDNA suggests the existence of other *cis*-acting elements in cold response.

These elements exist, as shown by the analysis of the promoter of the cold-responsive barley genes *blt4.9* and *blt101.1* (Dunn et al. 1998; Brown et al. 2001). In the *blt4.9* promoter, the hexanucleotide sequence CCGAAA was found to induce cold-responsive transcription activation (Dunn et al. 1998). This promoter element is very similar to the previously identified CCGACC-repeat. It was shown to bind nuclear proteins, but binding to CBFs was not confirmed. Interestingly, a conserved CCGAC sequence also present in this promoter and promoters of homologous genes did not bind nuclear proteins. Recently, a new cold-responsive element has been identified by Brown et al. (2001) when they found that the *blt101.1* promoter contains a 10-bp low temperature-response element (LTRE) with the sequence AAGAAGATGC. This element contributes to a two- to three-fold increase in transcriptional activity at low-temperature conditions compared to control conditions. The *blt101.1* promoter is not induced by high salt, drought or any other dehydration or stress-related conditions. This LTRE resembles a previously identified reversed salicylic acid-responsive TCA element (TCATCTTCTT) also found in barley (Goldsbrough et al. 1993).

Another class of well-studied drought/cold-responsive elements are the ABA-responsive elements (ABREs) (for review see Busk and Pagès 1998). These were originally identified already in 1990 when Mundy et al. (1990) and Guiltinan et al. (1990) found elements with the sequences TACGTGGC or GGACACGTGGC in the promoters of the rice *RAB* gene and the wheat *Em* gene, respectively. These first ABREs, which were found to be conserved and crucial to ABA response, are characterized by the presence of an ACGT-core sequence. Recently, the sequence requirement of these elements was determined to be ACGTGG/TC (Hattori et al. 2002), which fits very well with the strong ABRE sequence gACACGTGG/tC, to which the ABF1 (ABRE Binding Factor 1) protein was found to bind frequently (Choi et al. 2000). Closely related to these so-called G/ABREs are the C/ABREs like the CE3 element (Shen et al. 1996) or the motif III of rice (Ono et al. 1996), both containing a CGCGTG-core sequence (Choi et al. 2000). CE3 is a coupling element, like CE1 (TGCCACCGG; Shen and Ho 1995), which is often active in combination with G/ABREs (Shen et al. 1996).

In this same class of elements responding to ABA and dehydration it is worth mentioning a 67-bp fragment of the *Arabidopsis rd22* gene. This fragment contains two closely located MYC-like protein binding sites (CACATG) and a putative MYB-like protein binding site (TGGTTAG) of which the first MYC-element and the MYB-element were found to be essential for dehydration and ABA response (Abe et al. 1997). The second MYC element acts as a negative regulator of expression. A MYC-like bHLH protein named rd22BP1 and the ATMYB2 protein can both bind to the 67-bp *rd22* promoter fragment and act as transcriptional activators of the *rd22* gene in response to dehydration and ABA.

6.5 Ozone-Induced Gene Regulation

The air pollutant ozone has recently been found to resemble fungal elicitors by the induction of certain genes and biosynthetic pathways associated with pathogens (Sandermann et al. 1998). However, there is little information on ozone-responsive promoters. The best-characterized promoters are the resveratrol promoter of grapevine (*vst1*) (Schubert et al. 1997) and the class I *ß*-1,3,-glucanase promoters of tobacco (*GLB, Gln2*) (Grimmig et al. 2003). Analysis of transgenic tobacco plants, harboring 5'-deletion constructs of the *vst1* promoter fused to the bacterial *ß*-glucuronidase reporter gene, indicated that the DNA region of –430 to –280 bp is regulating ozone-induced stilbene synthase gene expression (Grimmig et al. 1997; Schubert et al. 1997). A search for potential *trans* factor-binding sites revealed several putative regulatory sequences of the ozone-responsive *vst1* promoter region (Grimmig et al. 1997; Schubert et al. 1997; Ernst et al. 1999).

MYB-like recognition elements have been found in the ozone-responsive *vst1* promoter region. Such elements have also been described for genes involved in pathogen defense (Rushton and Somssich 1998), in drought- or abscisic acid-responsive promoters (Abe et al. 1997). Furthermore, in defined boxes of genes of the phenylpropanoid pathway MYB elements are also present (Douglas 1996). As many of such genes are also induced by ozone it might be possible that MYB-like recognition elements also play a role in ozone-induced gene regulation.

G boxes are involved in gene regulation of diverse environmental factors like UV radiation, pathogen attack or wounding (Menkens et al. 1995), and two G box-like elements have also been described in the ozone-responsive *vst1* region (Ernst et al. 1999). As G boxes often function in concert with other *cis* elements, they may also be important during ozone stress, probably in combination with additional *cis* elements.

It is well known that exposure of plants to ozone results in the formation of stress ethylene (Pell et al. 1997; Sandermann et al. 1998), and it has been shown that the *vst1* promoter is induced by ethylene (Grimmig et al. 1997). A GCC box within the ethylene-responsive elements (ERE) is well conserved in several ethylene-inducible genes and represents the binding site for ERE-binding proteins (Leubner-Metzger et al. 1998; Yamamoto et al. 1999). It has been speculated that an inverse GCC box-like element present in the *vst1* promoter, which is disrupted in the ozone-insensitive –280 bp 5'-deletion construct, might be involved in ozone signalling (Schubert et al. 1997). However, an ozone-responsive region could be separated from an ethylene-responsive region (Grimmig et al. 2002). Furthermore, application of the ethylene response inhibitor 1-methylcyclopropene indicates that ethylene is not involved in signalling ozone-induced *vst1* gene regulation (Grimmig et al. 2003). In contrast, ozone induction of the tobacco class I β-1,3-glucanase and the basic type pathogenesis-related PR-1 protein promoter is mediated via ethylene. A functional GCC box was shown to be necessary for an ozone-induced regulation of these promoters (Fig. 6.3; Grimmig et al. 2003).

Two core sequences of elicitor-responsive W boxes ((T)TGACT) are present in the ozone-responsive *vst1* promoter (Schubert et al. 1997; Ernst et al. 1999). The presence of such W-box elements indicates that WRKY proteins might be active as transcription factors upon ozone fumigation, similarly as described upon elicitation (Eulgem et al. 2000). However, as the ozone-responsive region differs from the basal pathogen-responsive region (–280 to –140 bp; Schubert et al. 1997), additional *cis* elements must be involved in basal elicitation-inducible DNA binding. Using the ozone-responsive *vst1* region in the yeast one-hybrid system two *trans* factors, belonging to the WRKY proteins of group II, could be isolated (B. Heidenreich, D. Ernst, H. Sandermann; unpubl. results). Therefore, it might be speculated that WRKY genes are also ozone-inducible, as described for pathogen attack, elicitors, salicylic acid, wounding or mechanical stress (Eulgem et al. 2000). A schematic model for an ozone-induced activation of the *vst1* promoter by WRKY proteins and of *GLB* basic-type glucanase promoters by ethylene-responsive element-binding factors is shown in Fig. 6.4.

Apart from these two promoters, few data are available on other ozone-inducible promoters. The promoter of a senescence-associated gene (*SAG13*) isolated from *Arabidopsis thaliana* has been compared with the ozone-responsive region of the *vst1* promoter (Miller et al. 1999). A comparison of the *SAG13*

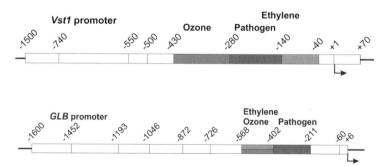

Fig. 6.3. Schematic representation of the *Vst1* promoter of the resveratrol synthase gene and the *GLB* promoter of the class I β-1,3-glucanase gene. *Ozone-, pathogen-, and ethylene-induced regions* are indicated

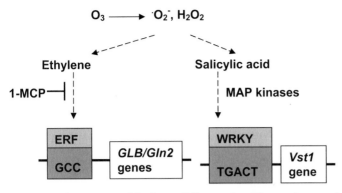

Fig. 6.4. Schematic model of two different signalling pathways of ozone-induced gene expression. The specific DNA sequence motifs for TF binding are indicated. *ERF* Ethylene-responsive element-binding factor; *WRKY* WRKY transcription factor; *1-MCP* 1-methylcyclopropene

promoter with the 150-bp ozone-responsive *vst1* region did not reveal any strong similarity (Miller et al. 1999). Similarly, comparison of a pinosylvin-*O*-methyltransferase promoter showed no strong similarity to the ozone-responsive *vst1* promoter region (Chiron et al. 2000). Sequence analysis of an ozone-inducible cytosolic superoxide dismutase gene from aspen revealed sequence elements similar to binding sites for transcription factors such as NF-κB and AP-1, which are known to be redox-regulated (Akkapeddi et al. 1999).

Taking all these sequence data and DNA motifs together, it is necessary to keep in mind that most of the results obtained are based on adequate databases and computer-based programs. An in vivo proof for the functionality of an "ozone-responsive" element, as has been shown for ethylene or elicitors, (e.g., Leubner-Metzger and Meins 1999; Eulgem et al. 2000), has not been carried out as yet.

6.6 Xenobiotic-Responsive cis Elements

Glutathione *S*-transferases (GSTs) are multifunctional enzymes that play important roles in the detoxification of xenobiotics and protect tissue against oxidative damage. *GST* transcripts are induced by a broad range of environmental factors such as pathogen attack, hormones, herbicides and also ozone. A well-characterized promoter sequence, the octopine synthase (*ocs*) element (ATCTTATGTCATTGATGACGACCTCC), has been shown to be important for GST expression (Chen et al. 1996; Chen and Singh 1999). Its function is a response to oxidative stress and detoxification of xenobiotics. *Ocs* elements first were found in the octopine (*ocs*) and nopaline (*nos*) synthase promoter of *Agrobacterium tumefaciens* (Yang et al. 1997). In the CaMV 35S promoter the *ocs* element is also referred to as the *as-1* site (cauliflower mosaic virus *activation sequence-1*) (Yang et al. 1997). Comparison of several *GST* promoters and these pathogen promoters revealed a core sequence of TGACG(T/C)AAG(CGA/GAC)T(G/T)ACG(TAA/CCC). In soybean the *GH2/4* gene encodes a GST, and it has been shown that this promoter is induced by plant hormones, H_2O_2 and also DTT (Ulmasov et al. 1995). *Trans* factors, binding to the *ocs* element (OBFs) have been isolated from several plants (Zhang et al. 1995; Chen et al. 1996). They belong to the basic region/leucine zipper motif transcription factors (bZIP). The bZIP proteins are involved in regulating processes upon defense, light and stress signalling, as well as seed maturation and flower development (Jakoby and Weishaar 2002). Referring to the *as-1* element the bZIP factor is called TGA1a (Johnson et al. 2001). TGA1a specifically binds *as-1* elements and xenobiotic stress enhances binding of TGA1a to *as-1*, thus activating transcription of the target gene (Klinedinst et al. 2000; Johnson et al. 2001).

Xenobiotic-responsive elements (XRE) are well characterized in animal gene promoters. The core sequence TnGCGTG or GCGTG is activated by aryl hydrocarbons (dioxin) (see http://www.genomatix.de). Computer analysis for various regulatory elements of chloroplastic superoxide dismutase *Sod1* gene in maize resulted in the identification of two XREs (Kernodle and Scandalios 2001). However, deletion analysis of the *Sod1* promoter to elucidate these XREs has not been carried out. Induction of *Sod1* may accelerate the neutralization of the superoxide anion and thus reduce the oxidative damage associated with aryl hydrocarbons toxicity (Cho et al. 2001).

6.7 Activation of DNA Sequences by Heavy Metal Ions

Heavy metals do not form a homogeneous group. Some heavy metals, like Fe, Zn, Ni and Cu, are essential elements, which a plant needs to grow and develop. Others, like Pb, Cd, Al, Hg or As, are non-essential, but are taken up by plants probably because of their chemical resemblance to essential elements. It is known that high or even elevated concentrations of heavy metals cause abiotic stress. Three types of stresses are generally discerned: production of reactive oxygen species because of auto-oxidation, typically in the case of Fe and Cu; blocking of essential functional groups in biomolecules, particularly for Hg and Cd; displacement of essential metal ions from biomolecules (Schützendübel and Polle 2002). Although oxidative stresses are only expected upon exposure to Fe and Cu, the redox-active heavy metals, also exposure to Cd, Zn, Ni and Al causes oxidative stress-related symptoms as a consequence of their interaction with anti-oxidative systems (Madhava Rao and Sresty 2000; Schützendübel et al. 2001; Yamamoto et al. 2002).

Relatively little is known about the transcriptional activation of plant genes in response to heavy metal ion exposure. Exposure to metals that are normally detoxified by binding to phytochelatins induces the expression of genes involved in phytochelatin synthesis and its gluthatione-based precursors such as the *gsh1*, *gsh2* and *gr1* genes of *Arabidopsis*, respectively encoding γ-EC synthetase, GSH synthethase and GSH reductase (Xiang and Oliver 1998). Metal response elements (MREs) have been identified in animals (TGCRCNC) that are recognized by MRE binding transcription factors (MTFs) in regulating the response to (excess) heavy metals like Zn, Cd and Cu (e.g., Giedroc et al. 2001; Zhang et al. 2001a). For yeast a similar system has been described where the Zap1p transcriptional activator controls expression of Zn-responsive genes through binding to the Zn-responsive element (ACCYYNAAGGT) (Lyons et al. 2000).

In plants these metal response elements have not been found. Recently, the transcriptional response of *Arabidopsis* to Cd and long-term Hg exposure has been studied. For Cd, many of the induced genes could also be induced by other stresses, including oxidative stress (Suzuki et al. 2001). For Hg, seven genes (out of 576 clones analyzed) were found to be exclusively Hg-induced, but promoter elements were not studied (Heidenreich et al. 2001). The same holds true for the PvSR2 gene from bean (*Phaseolus vulgaris*), of which the expression is strongly induced by heavy metals (Hg, Cd, As, Cu), but not by UV, temperature or pathogen stress (Zhang et al. 2001b). The only DNA element with some relation to metal response is the iron-dependent regulatory sequence (IDRS, CACGAGNCCNCCAC), which is responsible for the iron-regulated transcriptional repression of plant ferritin genes (Petit et al. 2001).

6.8 Elicitor-Responsive Elements

The focus here is on elicitation of plants by pathogens and not by abiotic elicitors such as heavy metals, ozone or xenobiotics. There are many *cis* elements known to be involved in signalling pathogen attack and elicitation of plants, and there is cross-talk between the different induced response pathways (Maleck and Dietrich 1999). Plant hormones like ethylene, salicylic acid or jasmonic acid are involved, and specific *cis* and regulatory elements have been described for these substances (Durner et al. 1997; Leubner-Metzger and Meins 1999; Memelink et al. 2001). In addition, G boxes, the bZIP factors as well as MREs and MYB proteins interact in a fine-tuned gene regulation upon pathogen attack. An excellent review about *cis*-acting elements within regulatory regions of pathogen-responsive promoters and the corresponding *trans* factors is given by Rushton and Somssich (1998).

AT-rich sequences are also known to be involved in the activation of *CHS* in pea upon treatment with fungal elicitor (Seki et al. 1996). Recently, a novel pathogen-responsive, AT-rich element in the promoter of a tobacco gene involved in early steps of an incompatible plant/pathogen interaction has been described (Pontier et al. 2001). This hypersensitive response-responsive *cis* element (HSRE) consists of a 12-bp motif (TAAAATTCTTTG) and shows no homologies to known regulatory regions (Pontier et al. 2001). However, in members of the *Eli7* gene family of parsley, which are rapidly activated by fungal elicitor or pathogen infection, derivatives of the HSRE motif are present (Kirsch et al. 2000). This indicates that HSR-like elements are important for induction of genes preferentially expressed during the hypersensitive response.

Elictor-responsive elements (W boxes) and structurally related cognate DNA-binding proteins have been characterized in detail (Eulgem et al. 1999, 2000; Chen and Chen 2000). The W-box elements contain a core motif of (T)(T)TGAC(C/T) and are arranged as single-sequence elements, tandems or palindromes (Eulgem et al. 1999; Yang et al. 1999). The binding *trans* factors belong to the WRKY proteins, a superfamily of transcription factors, and in *Arabidopsis* about 100 members have been described (Chen and Chen 2000; Eulgem et al. 2000). A common feature is a highly conserved 60-amino acid stretch with the strongly conserved WRKYGQK peptide. WRKY factors possess a potential zinc-finger motif and have been classified into three groups (Eulgem et al. 2000). Like other transcription factors the WRKY proteins are targeted to the nucleus (Eulgem et al. 1999; Hara et al. 2000; Robatzek and Somssich 2001). WRKY factors bind to a variety of defense-related genes, including the *WRKY* genes themselves (Rushton and Somssich 1998; Eulgem et al. 1999; Yang et al. 1999). Thus, an autoregulatory binding to its own promoter may occur (Eulgem et al. 1999). A new type of W box-containing elicitor-responsive element has been identified in parsley (Yang et al. 1998; Kirsch et al. 2001). The central GTCA-containing region impaired elicitor responsiveness. Furthermore, a tandem repeat with the well-known W box (TGAC/GTCA) seems to be of relevance for a rapid elicitor response (Kirsch et al. 2001). Similar to the *WRKY* genes, the *Eli17* gene encodes a protein with structural characteristics of transcription factors (Kirsch et al. 2001). The diversity of binding sites and of putative transcription factors indicates a subtle and complex regulation of numerous genes involved in elicitor-induced gene regulation, similar to what has been reported for a transcriptional reprogramming of primary and secondary metabolism in elicitated parsley cells (Batz et al. 1998).

6.9 Plant Hormone-Responsive DNA Sequences (Auxin, Cytokinin and Ethylene)

The plant hormones auxin, cytokinin, ethylene and abscisic acid are generally associated with plant growth and development, but they also play a role in stress response. The action of abscisic acid in relation to stress has already been described in section 6.4. This section deals with the response to auxin, cytokinin and ethylene in relation to stress.

Next to its role as a plant growth regulator, auxin is known to repress the transcription of defense-related genes (Jouanneau et al. 1991). Probably the best-studied auxin response element is the *ocs*-element from *Agrobacterium tumefaciens*, similar to the *as-1* element from the CaMV 35S promoter (Bouchez et al. 1989), responding to both auxin and SA (see section 6.10) and also found in the promoter of GST genes (see section 6.6). However, in general the regulation of transcription by auxin is mediated by the presence of the auxin-response element (AuxRE), a TGTCTC-sequence recognized by the Auxin Response Factor 1 (ARF1) as cloned from *Arabidopsis* (Ulmasov et al. 1997). Heterodimerization of ARFs with AUX/IAA proteins controls the response of AuxRE-containing promoters to auxin (Tiwari et al. 2001).

Cytokinin generally antagonizes the effect of auxin. Recently, great advances have been made in understanding more about the cytokinin-signalling pathway (Sheen 2002), but the information on *cis* elements regulating cytokinin response is still sparse. Transcription of the *Arabidopsis* response regulator genes *ARR4, ARR5, ARR6, ARR7, ARR15* and *ARR16* increases rapidly upon induction by cytokinin (D'Agostino et al. 2000). This response is also seen for the 1.6-kb *ARR5* promoter when fused to GUS, but *cis*-acting elements have not yet been identified.

Ethylene influences processes in plant growth and development such as fruit ripening and leaf senescence, but also in the responses to biotic and abiotic stresses. Sequences controlling developmental ethylene response appear to be different from the ethylene-responsive *cis* elements found in defense-related genes (Ohme-Takagi et al. 2000). Promoters of many defense-related genes contain a so-called GCC box (Ohme-Takagi and Shinshi 1995; Sato et al. 1996). This ethylene-responsive element (TAAGAGCCGCC with a GCCGCC-core) is recognized by ethylene-responsive element binding factors (ERFs) such as the AtERF1 to AtERF5 proteins identified in *Arabidopsis* (Fujimoto et al. 2000). These proteins can act both as transcriptional activators and repressors of upstream genes.

6.10 Salicylic Acid and Jasmonate *cis*-Acting Elements

Salicylic acid (SA) and jasmonic acid (JA) (or derivatives like jasmonate or methyl-jasmonate) both have a role in plant defense against pathogens (reviewed by Vernooij et al. 1994 and Reymond and Farmer 1998). Nevertheless, these defense pathways appear to be separate (Thomma et al. 1998), with only less than a quarter of the genes for which transcription responds to either of these substances (Schenk et al. 2000). Therefore, their respective response elements will not be found very often together in one promoter.

SA was first found to activate genes containing the TCA box, a 10-bp element (TCATCTTCTT) originally identified in the barley β-1,3-glucanase gene, but then found in many more SA-responsive genes. This element is recognized by the tobacco TCA-1 nuclear protein (Goldsbrough et al. 1993). SA is also able to induce the activity of promoters carrying the *ocs*-element or *as-1* box (Zhang and Singh 1994), the 20-bp element [TGACG(T/C)AAG (C/G)(G/A)(A/C)T(G/T)ACG(T/C) (A/C)(A/C)] originally identified in the CaMV promoter and in *Agrobacterium tumefaciens* T-DNA (Bouchez et al. 1989) and discussed previously in this chapter. Linker scanning mutagenesis of the PR-1 promoter of *Arabidopsis* revealed another motif, GGGACTTTTCC, with sequence similarity to the animal κB sequence, to be important for response to INA, another inducer of systemic acquired resistance (SAR) like SA (Lebel et al. 1998). SA is also able to induce the transcription of genes containing WRKY boxes (based on a TGAC-core sequence) in their promoters through the activation of WRKY domain-containing DNA-binding factors such as the tobacco TDBA12 protein (Yang et al. 1999).

Jasmonates like JA or methyl jasmonate (MeJA) are lipid-derived signalling molecules regulating metabolic processes, reproduction and defense- or injury-related responses in plants (Liechti and Farmer 2002). The studies of promoter elements directing the response to JA are not as advanced as for SA. One of the JA-responsive genes, which has been studied in some detail, is the PDF1.2 defensin gene of *Arabidopsis*, which is clearly induced by MeJA and not by SA. The promoter of this gene contains several putative promoter elements of which the combination of a GCC box subtending a G box has been seen in other PR-gene promoters (Manners et al. 1998). Previously, a G box (TCACGTGG) was found to be essential for MeJA-responsiveness of the potato PI-II promoter (Kim et al. 1992), and together with a C-rich region (CCCTATAGGG), it has been suggested as a putative MeJA box in the peroxidase Shpx6-gene of *Stylosanthes humilis* (Curtis et al. 1997). A GCC box essential for JA-responsiveness was also identified in the upstream region of the strictosidine synthase gene of *Catharanthus roseus* (Menke et al. 1999), where it was recognized by the AP2-domain-containing ORCA2 transcription factor. Recently, two other JA-response elements, JASE1 (CGTCAATGAA) and JASE2

(CATACGTCGTCAA), were found in the promoter of the *OPR1* gene of *Arabidopsis* (He and Gan 2001). JASE2 resembles mixed A/C boxes (ACGT-core), generally recognized by bZIP transcription factors, but JASE1 appears to be novel.

6.11 Concluding Remarks

The number of known *cis* elements and cognate DNA-binding factors involved in stress-induced gene regulation has increased during the last years and is still rising. In terms of ecotoxicology, several stressors may act at the same time, thus resulting in a complex interaction of several *cis/trans* factors. Plant genes contain a set of *cis*-acting elements, which may contribute, more or less, to a balanced gene expression. Recently, it was shown that in parsley the pathogen defense overrides UV protection through an inversely regulated ACGT/ACGT promoter element of a light-responsive promoter element (Logemann and Hahlbrock 2002). Expression profiling using microarray technologies will identify transcripts, which are induced by the same external stimuli. The completion of the *Arabidopsis* genome sequence and ongoing plant genome projects will contribute to the identification of plant promoters. Database-assisted promoter analyses will result in the detection of patterns in such promoters, leading to the identification of transcription factor-binding sites (Harmer et al. 2000). The construction of synthetic plant promoters containing defined *cis* elements will provide novel insights into the interaction of ecotoxicologically important stress factors (Rushton et al. 2002). With the completion of plant genome projects (*Arabidopsis*, rice), the full complement of genes coding for transcription factors from these plants can be identified and described (Riechmann et al. 2000). In combination with a genome-wide mRNA-profiling, the complex interaction of transcription factors can be analyzed, thus unravelling networks in plant gene expression upon environmental stress, similar to what has been reported for flavonoid pathway genes, MYB transcription factor C1 and helix-loop-helix protein R (Bruce et al. 2000). Recently, Aharoni and Vorst (2001) wrote an excellent review on DNA microarray technology, metabolic pathways, environmental stress responses and regulatory networks. Thus, effective and efficient resources are available to identify diverse transcriptional regulators.

A global analysis of genome-wide expression data in yeast has been published (Ihmels et al. 2002). Using all available expression data on yeast, regulatory properties of cellular pathways and *cis* elements were characterized. In addition a comprehensive set of overlapping transcriptional modules was identified. A similar approach using expression data on *Arabidopsis thaliana* should provide more insights into the orchestrated regulated transcriptional network upon abiotic/biotic stress. Further studies, combining receptors, signalling cascades, transcription factors and response elements, as well as in silico analysis of modules and transcriptional networks, will provide a better understanding of plant fitness in altering ecosystems.

Acknowledgements. This work has been supported by BayForUV (Bayerisches Staatsministerium für Wissenschaft, Forschung und Kunst), EUROSILVA (BMBF) and Deutsche Forschungsgemeinschaft (SFB 607). Thanks are also due to the following coworkers: Evi Bieber, Hélène Chiron, Maria Nuria Gonzalez-Perez, Bernhard Grimmig, Bernd Heidenreich and Roland Schubert.

References

Abe H, Yamaguchi-Shinozaki K, Urao T, Iwasaki T, Hosokawa D, Shinozaki K (1997) Role of Arabidopsis MYC and MYB homologs in drought- and abscisic acid-regulated gene expression. Plant Cell 9:1859–1868

Aharoni A, Vorst O (2001) DNA microarrays for functional plant genomics. Plant Mol Biol 48:99–118

Almoguera C, Prieto-Dapena P, Jordano J (1998) Dual regulation of a heat shock promoter during embryogenesis: stage-dependent role of heat shock elements. Plant J 13:437–446

Akkapeddi AS, Noormets A, Deo BK, Karnosky DF, Podila GK (1999) Gene structure and expression of the aspen cytosolic copper/zinc-superoxide dismutase (PtSodCc1). Plant Sci 143:151–162

Ausubel FM, Brent R, Kingston RE, Moore DD, Seidman JG, Smith JA, Struhl K (2001) Current protocols in molecular biology. Wiley, New York

Baker SS, Wilhelm KS, Tomashow MF (1994) The 5'-region of *Arabidopsis thaliana cor15l* has *cis*-acting elements that confer cold-, drought- and ABA-regulated gene expression. Plant Mol Biol 24:701–713

Batz O, Logemann E, Reinold S, Hahlbrock K (1998) Extensive reprogramming of primary and secondary metabolism by fungal elicitor or infection in parsley cells. Biol Chem 379:1127–1135

Baxevanis AD (2001) The molecular biology database collection: an updated compilation of biological database resources. Nucleic Acids Res 29:1–10

Bharti K, Schmidt E, Lyck R, Heerklotz D, Bublak D, Scharf K-D (2000) Isolation and characterization of HsfA3, a new heat stress transcription factor of *Lycopersicon peruvianum*. Plant J 22:355–366

Bouchez D, Tokuhisa JG, Llewellyn DJ, Dennis ES, Ellis JG (1989) The ocs-element is a component of the promoters of several T-DNA and plant viral genes. EMBO J 8:4197–4204

Bray EA, Bailey-Serres J, Weretilnyk E (2000) Responses to abiotic stress. In: Buchanan BB, Gruissem W, Jones RL (eds) Biochemistry and molecular biology of plants. American Society of Plant Physiologists, Rockville, pp 1158–1203

Brosché M, Fant C, Bergkvist SW, Strid H, Svensk A, Olsson O, Strid Å (1999) Molecular markers for UV-B stress in plants: alteration of the expression of four classes of genes in *Pisum sativum* and the formation of high molecular mass RNA adducts. Biochim Biophys Acta 1447:185–198

Brown APC, Dunn MA, Goddard NJ, Hughes MA (2001) Identification of a novel low-temperature-response element in the promoter of the barley (*Hordeum vulgare* L) gene *blt101.1*. Planta 213:770–780

Bruce W, Folkerts O, Garnaat C, Crasta O, Roth B, Bowen B (2000) Expression profiling of the maize flavonoid pathway genes controlled by estradiol-inducible transcription factors CRC and P. Plant Cell 12:65–79

Busk PK, Pagès M (1998) Regulation of abscisic acid-induced transcription. Plant Mol Biol 37:425–435

Chen C, Chen Z (2000) Isolation and characterization of two pathogen- and salicylic acid-induced genes encoding WRKY DNA-binding proteins from tobacco. Plant Mol Biol 42:387–396

Chen W, Singh KB (1999) The auxin, hydrogen peroxide and salicylic acid induced expression of the *Arabidopsis GST6* promoter is mediated in part by an ocs element. Plant J 19:667–677

Chen W, Chao G, Singh KB (1996) The promoter of a H_2O_2-inducible, *Arabidopsis* glutathione S-transferase contains closely linked OBF- and OBP1-binding sites. Plant J 10:955–966

Chen W, Provart NJ, Glazebrook J, Katagiri F, Chang HS, Eulgem T, Mauch F, Luan S, Zou G, Whitham SA, Budworth PR, Tao Y, Xie Z, Chen X, Lam S, Kreps JA, Harper JF, Si-Ammour A, Mauch-Mani B, Heinlein M, Kobayashi K, Hohn T, Dangl JL, Wang X, Zhu T (2002) Expression profile matrix of Arabidopsis transcription factor genes suggests their putative functions in response to environmental stresses. Plant Cell 14:559–574

Chiron H, Drouet A, Lieutier F, Payer H-D, Ernst D, Sandermann H (2000) Gene induction of stilbene biosynthesis in Scots pine in response to ozone treatment, wounding, and fungal infection. Plant Physiol 124:865–872

Cho JS, Chang MS, Rho HM (2001) Transcriptional activatuion of the human Cu/Zn superoxide dismutase gene by 2,3,7,8-tetrachlordibenzo-p-dioxin through the xenobiotic-responsive element. Mol Gen Genomics 266:133–141

Choi H-I, Hong J-H, Ha J-O, Kang J-J, Kim SY (2000) ABFs, a family of ABA-responsive element binding factors. J Biol Chem 275:1723–1730

Curtis MD, Rae, AL, Rusu AG, Harrison SJ, Manners JM (1997) A peroxidase gene promoter induced by phytopathogens and methyl jasmonate in transgenic plants. Mol Plant Microbe Interact 10:326–338

Czarnecka-Verner E, Yuan C-X, Scharf K-D, English G, Gurley WB (2000) Plants contain a novel multi-member class of heat shock factors without transcriptional activator potential. Plant Mol Biol 43:459–471

D'Agostino IB, Deruere J, Kieber JJ (2000) Characterization of the response of the Arabidopsis response regulator gene family to cytokinin. Plant Physiol 124:1706–1717

Day TA (1993) Relating UV-B radiation screening effectiveness of foliage to absorbing-compound concentration and anatomical characteristics in a diverse group of plants. Oecologia 95:542–550

Douglas CJ (1996) Phenylpropanoid metabolism and lignin biosynthesis: from weeds to trees. Trends Plant Sci 1:171–178

Dunn MA, White AJ, Vural S, Hughes MA (1998) Identification of promoter elements in a low-temperature-responsive gene (*blt4.9*) from barley (*Hordeum vulgare* L.). Plant Mol Biol 38:551–564

Durner J, Shah J, Klessig DF (1997) Salicylic acid and disease resistance in plants. Trends Plant Sci 2:266–274

Ernst D, Grimmig B, Heidenreich B, Schubert R, Sandermann H (1999) Ozone-induced genes: mechanisms and biotechnological applications. In: Smallwood MF, Calvert

CM, Bowles DJ (eds) Plant responses to environmental stress. BIOS Scientific Publishers, Oxford, pp 33–41

Eulgem T, Rushton PJ, Schmelzer E, Hahlbrock K, Somssich IE (1999) Early nuclear events in plant defence signalling: rapid gene activation by WRKY transcription factors. EMBO J 18:4689–4699

Eulgem T, Rushton PJ, Robatzek S, Somssich IE (2000) The WRKY superfamily of plant transcription factors. Trends Plant Sci 5:199–206

Ferl R, Paul A-L (2000) Genome organization and expression. In: Buchanan BB, Gruissem W, Jones RL (eds) Biochemistry and molecular biology of plants. American Society of Plant Physiologists, Rockville, pp 312–357

Foster R, Izawa T, Chua N-H (1994) Plant bZIP proteins gather at ACGT elements. FASEB J 8:192–200

Fujimoto SY, Ohta M, Usui A, Shinshi H, Ohme-Takagi M (2000) Arabidopsis ethylene-responsive element binding factors act as transcriptional activators or repressors of GCC box-mediated gene expression. Plant Cell 12:393–404

Giedroc DP, Chen X, Apuy JL (2001) Metal response element (MRE)-binding transcription factor-1 (MTF-1): structure, function, and regulation. Antioxid Redox Signal 3:577–596

Giuliano G, Pichersky E, Malik VS, Timko MP, Scolnik PA, Cashmore AR (1988) An evolutionarily conserved protein binding sequence upstream of a plant light-regulated gene. Proc Natl Acad Sci USA 85:7089–7093

Goldsbrough AP, Albrecht H, Stratford R (1993) Salicylic acid-inducible binding of a tobacco nuclear protein to a 10 bp sequence which is highly conserved amongst stress-inducible genes. Plant J 3:563–571

Green PJ, Yong M-H, Cuozzo M, Kano-Murakimi Y, Silverstein P, Chua N-H (1988) Binding site requirements for pea nuclear protein factor GT-1 correlate with sequences required for light-dependent transcriptional activation of the *rbcS-3A* gene. EMBO J 7:4035–4044

Grimmig B, Schubert R, Fischer R, Hain R, Schreier PH, Betz C, Langebartels C, Ernst D, Sandermann H (1997) Ozone- and ethylene-induced regulation of a grapevine resveratrol synthase promoter in transgenic tobacco. Acta Physiol Plant 19:467–474

Grimmig B, Gonzalez-Perez MN, Welzl G, Penuelas J, Schubert R, Hain R, Heidenreich B, Betz C, Langebartels C, Ernst D, Sandermann H (2002) Ethylen- and ozone-induced regulation of a grapevine resveratrol synthase gene: different responsive promoter regions. Plant Physiol Biochem 40:865–870

Grimmig B, Gonzalez-Perez MN, Leubner-Metzger G, Vögeli-Lange R, Meins F, Hain Penuelas J, Heidenreich B, Langebartels C, Ernst D, Sandermann H (2003) Ozone-induced gene expression occurs via ethylene-dependent and -independent signalling. Plant Mol Biol 51:599–607

Guiltinan MJ, Marcotte WR, Quatrano RS (1990) A plant leucine zipper protein that recognizes an abscisic acid response element. Science 250:267–271

Hara K, Yagi M, Kusano T, Sano H (2000) Rapid systemic accumulation of transcripts encoding tobacco WRKY transcription factor upon wounding. Mol Gen Genet 263:30–37

Harmer SL, Hogenesch JB, Straume M, Chang H-S, Han B, Zhu T, Wang X, Kreps JA, Kay SA (2000) Orchestrated transcription of key pathways in *Arabidopsis* by the circadian clock. Science 290:2110–2113

Hartmann U, Valentine WJ, Christie JM, Hays J, Jenkins GI, Weisshaar B (1998) Identification of UV/blue light-response elements in the *Arabidopsis thaliana* chalcone synthase promoter using a homologous protoplast transient expression system. Plant Mol Biol 36:741–754

Hattori T, Totsuku M, Hobo T, Kagaya Y, Yamamoto-Toyoda A (2002) Experimentally determined sequence requirements of ACGT-containing abscisic acid response element. Plant Cell Physiol 43:136–140

He Y, Gan S (2001) Identical promoter elements are involved in regulation of the *OPR1* gene by senescence and jasmonic acid in *Arabidopsis*. Plant Mol Biol 47:595–605

Hehl R, Wingender E (2001) Database-assisted promoter analysis. Trends Plant Sci 6:251–255

Heidenreich B, Mayer K, Sandermann H, Ernst D (2001) Mercury-induced genes in *Arabidopsis thaliana*: identification of induced genes upon long-term mercuric ion exposure. Plant Cell Environ 24:1227–1234

Hemm MR, Herrmann KM, Chapple C (2001) AtMYB4: a transcription factor general in the battle against UV. Trends Plant Sci 4:135–136

Higo K, Ugawa Y, Iwamoto M, Korenaga T (1999) Plant cis-acting regulatory DNA elements (PLACE) database: 1999. Nucleic Acids Res 27:297–300

Hughes MA (1996) Plant molecular genetics. Longman, Essex

Ihmels J, Friedlander G, Bergmann S, Sarig O, Ziv Y, Barkai N (2002) Revealing modular organization in the yeast transcriptional network. Nat Genet 31:370–377

Jakoby M, Weisshaar B (2002) bZIP transcription factors in *Arabidopsis*. Trends Plant Sci 7:106–111

Jensen LJ, Knudsen S (2000) Automatic discovery of regulatory patterns in promoter regions based on whole cell expression data and functional annotation. Bioinformatics 16:326–333

Jin H, Cominelli E, Bailey P, Parr A, Mehrtens F, Jones J, Tonelli C, Weisshaar B, Martin C (2000) Transcriptional repression by AtMYB4 controls production of UV-protecting sunscreens in *Arabidopsis*. EMBO J 19:6150–6161

Johnson C, Glover G, Arias J (2001) Regulation of DNA binding and *trans*-activation by a xenobiotic stress-activated plant transcription factor. J Biol Chem 276:172–178

Jordan BR (1996) The effects of ultraviolet-B radiation on plants: a molecular perspective. In: Callow JA (ed) Advances in botanical research, vol 22. Academic Press, San Diego, pp 97–162

Jouanneau J-P, Lapous D, Guern J (1991) In plant protoplasts, the spontaneous expression of defense reactions and the responsiveness to exogenous elicitors are under auxin control. Plant Physiol 96:459–466

Kernodle SP, Scandalios JG (2001) Structural organization, regulation, and expression of the chloroplastic superoxide dismutase *Sod1* gene in maize. Arch Biochem Biophys 391:137–147

Kim SR, Choi JL, Costa MA, An G (1992) Identification of G-box sequence as an essential element for methyl jasmonate response of potato proteinase inhibitor II promoter. Plant Physiol 99:627–631

Kirsch C, Takayma-Wik M, Schmelzer E, Hahlbrock K, Somssich IE (2000) A novel regulatory element invovlded in rapid activation of parsley *ELI7* gene family members by fungal elicitor or pathogen infection. Mol Plant Pathol 1:243–251

Kirsch C, Logemann E, Lippok B, Schmelzer E, Hahlbrock K (2001) A highly specific pathogen-responsive promoter element from the immediate-early activated *CMPG1* gene in *Petroselinum crispum*. Plant J 26:217–227

Klinedinst S, Pascuzzi P, Redman J, Desai M, Aria J (2000) A xenobiotic-stress-activated transcription factor and its cognate target genes are preferentially expressed in root tip meristems. Plant Mol Biol 42:679–688

Koch MA, Weisshaar B, Kroymann J, Haubold B, Mitchell-Olds T (2001) Comparartive genomics and regulatory evolution: conservation and function of the *Chs* and *Apetala3* promoters. Mol Biol Evol 18:1882–1891

Langebartels C, Schraudner M, Heller W, Ernst D, Sandermann H (2002) Oxidative stress and defense reactions in plants exposed to air pollutants and UV-B radiation. In: Inzé D, Van Montagu M (eds) Oxidative stress in plants. Taylor and Francis, London, pp 105–135

Lebel E, Heifetz P, Thorne L, Uknes S, Ryals J, Ward E (1998) Functional analysis of regulatory sequences controlling PR-1 gene expression in *Arabidopsis*. Plant J 16:223–233

Leubner-Metzger G, Meins F (1999) Functions and regulation of plant ß-1,3-glucanases (PR-2). In: Datta SK, Muthukrishnan S (eds) Pathogenesis-related proteins in plants. CRC Press, Boca Raton, pp 49–76

Leubner-Metzger G, Petruzzelli L, Waldvogel R, Vögeli-Lange R, Meins F (1998) Ethylene-responsive element binding protein (EREBP) expression and the transcriptional regulation of class I β-1,3-glucanase during tobacco seed germination. Plant Mol Biol 38:785–795.

Liechti R, Farmer EE (2002) The jasmonate pathway. Science 296:1649–1650

Liu Q, Kasuga M, Sakuma Y, Abe H, Miura S, Yamaguchi-Shinozaki K, Shinozaki K (1998) Two transcription factors, DREB1 and DREB2, with an EREBP/AP2 DNA binding domain separate two cellular signal transduction pathways in drought- and low-temperature-responsive gene expression, respectively, in *Arabidopsis*. Plant Cell 10:1391–1406

Logemann E, Hahlbrock K (2002) Crosstalk among stress responses in plants: pathogen defense overrides UV protection through an inversely regulated ACE/ACE type of light-responsive gene promoter unit. Proc Natl Acad Sci USA 99:2428–2432

Lyons TJ, Gasch AP, Gaither LA, Botstein D, Brown PO, Eide DJ (2000) Genome-wide characterization of the Zap1p zinc-responsive regulon in yeast. Proc Natl Acad Sci USA 97:7957–7962

Madhava Rao KV, Sresty TV (2000) Antioxidative parameters in the seedlings of pigeon-pea (*Cajanus cajan* (L.) Millspaugh) in response to Zn and Ni stresses. Plant Sci 157:113–118

Maleck K, Dietrich RA (1999) Defense on multiple fronts: how do plants cope with diverse enemies? Trends Plant Sci 4:215–219

Maleck K, Levine A, Eulgem T, Morgan A, Schmid J, Lawton KA, Dangl JL, Dietrich RA (2000) The transcriptome of *Arabidopsis thaliana* during systemic acquired resistance. Nat Genet 26:403–410

Manners JM, Penninckx IAMA, Vermaere K, Kazan K, Brown RL, Morgan A, Maclean DJ, Curtis MD, Cammue BPA, Broekaert WF (1998) The promoter of the plant defensin gene *PDF1.2* from *Arabidopsis* is sytemically activated by fungal pathogens and responds to methyl jasmonate but not to salicylic acid. Plant Mol Biol 38:1071–1080

Memelink J, Verpoorte R, Kijne JW (2001) ORCAnization of jasmonate-responsive gene expression in alkaloid metabolism. Trends Plant Sci 6:212–219

Menke FLH, Champion A, Kijne J, Memelink J (1999) A novel jasmonate- and elicitor-responsive element in the periwinkle secondary metabolite biosynthetic gene *Str* interacts with a jasmonate- and elicitor-inducible AP2-domain transcription factor, ORCA2. EMBO J 18:4455–4463

Menkens AE, Schindler U, Cashmore AR (1995) The G box: a ubiquitous regulatory DNA element in plants bound by the GBF family of bZIP proteins. Trends Biochem Sci 20:506–510

Miller JD, Arteca RN, Pell EJ (1999) Senescence-associated gene expression during ozone-induced leaf senescence in *Arabidopsis*. Plant Physiol 120:1015–1023

Mundy J, Yamaguchi-Shinozaki K, Chua N-H (1990) Nuclear proteins bind conserved elements in the abscisic acid-responsive promoter of a rice *rab* gene. Proc Natl Acad Sci USA 87:1406–1410

Nover L (1987) Expression of heat stress genes in homologous and heterologous systems. Enzyme Microb Technol 9:130–144

Nover L, Bharti K, Döring P, Mishra SK, Ganguli A, Scharf K-D (2001) *Arabidopsis* and the heat stress transcription factor world: how many heat stress transcription factors do we need? Cell Stress Chaperones 6:177–189

Ohme-Takagi M, Shinshi H (1995) Ethylene-inducible DNA binding proteins that interact with an ethylene-responsive element. Plant Cell 7:173–182

Ohme-Takagi M, Suzuki K, Shinshi H (2000) Regulation of ethylene-induced transcription of defense genes. Plant Cell Physiol 41:1187–1192

Ono A, Izawa T, Chua N-H, Shimamoto K (1996) The *rab16B* promoter of rice contains two distinct abscisic acid-responsive elements. Plant Physiol 112:483–491

Ouwerkerk PBF, Trimborn TO, Hilliou F, Memelink J (1999) Nuclear factors GT-1 and 3AF1 interact with multiple sequences within the promoter of the *Tdc* gene from Madagascar periwinkle: GT-1 is involved in UV light-induced expression. Mol Gen Genet 261:610–622

Pell EJ, Schlagnhaufer CD, Arteca RN (1997) Ozone-induced oxidative stress: mechanisms of action and reaction. Physiol Plant 100:264–273

Périer RC, Praz V, Junier T, Bonnard C, Bucher P (2000) The eukaryotic promoter database (EPD.) Nucleic Acids Res 28:302–303

Petit J-M, van Wuytswinkel O, Briat J-F, Lobréaux S (2001) Characterization of an iron-dependent regulatory sequence involved in the transcriptional control of *AtFer1* and *ZmFer1* plant ferritin genes by iron. J Biol Chem 276:5584–5590

Pontier D, Balagué C, Bezombes-Marion I, Tronchet M, Deslandes L, Roby D (2001) Identification of a novel pathogen-responsive element in the promoter of the tobacco gene *HSR203 J*, a molecular marker of the hypersensitive response. Plant J 26:495–507

Quandt K, Frech K, Karas H, Wingender E, Werner T (1995) MatInd and MatInspector: new fast and versatile tools for detection of consensus matches in nucleotide sequence data. Nucleic Acids Res 23:4878–4884

Reindl A, Schöffl F (1998) Interaction between the *Arabidopsis thaliana* heat shock transcription factor HSF1 and the TATA binding protein TBP. FEBS Lett 436:318–322

Reymond P, Farmer EE (1998) Jasmonate and salicylate as global signals for defense gene expression. Curr Opin Plant Biol 1:404–411

Riechmann JL, Heard J, Martin G, Reuber L, Jiang C-Z, Keddie J, Adam L, Pineda O, Ratcliffe OJ, Samah RR, Creelman R, Pilgrim M, Broun P, Zhang JZ, Ghandehari D, Sherman BK, Yu G-L (2000) *Arabidopsis* transcription factors: genome-wide comparative analysis among eukaryotes. Science 290:2105–2110

Robatzek S, Somssich IE (2001) A new member of the *Arabidopsis* WRKY transcription factor family, AtWRKY6, is associated with both senescence- and defence-related processes. Plant J 28:123–133

Rombauts S, Déhais P, Van Montagu M, Rouzé P (1999) Plant CARE, a plant *cis*-acting regulatory element database. Nucleic Acids Res 27:295–296

Rushton PJ, Somssich IE (1998) Transcriptional control of plant genes responsive to pathogens. Curr Opin Plant Biol 1:311–315

Rushton PJ, Reinstädler A, Lipka V, Lippok B, Somssich IE (2002) Synthetic plant promoters containing defined regulatory elements provide novel insights into pathogen- and wound-induced signalling. Plant Cell 14:749–762

Sakamoto A, Tanaka A, Watanabe H, Tango S (1998) Molecular cloning of *Arabidopsis* photolyase gene (*PHR1*) and characterization of its promoter region. DNA Sequence 9:335–340

Sambrook J, Russell DW (2001) Molecular cloning a laboratory manual. Cold Spring Harbor Laboratory Press, New York

Sandermann H, Ernst D, Heller W, Langebartels C (1998) Ozone: an abiotic elicitor of plant defence reactions. Trends Plant Sci 3:47–50

Sato F, Kitajima S, Koyama T, Yamada Y (1996) Ethylene-induced gene expression of osmotin-like protein, a neutral isoform of tobacco PR-5, is mediated by the AGC-CGCC cis-sequence. Plant Cell Physiol 37:249–255

Scharf K-D, Heider H, Höhfeld I, Lyck R, Schmidt E, Nover L (1998) The tomato Hsf system: HsfA2 needs interaction with HsfA1 for efficient nuclear import and may be localized in cytoplasmic heat stress granules. Mol Cell Biol 18:2240–2251

Schenk PM, Kazan K, Wilson I, Anderson JP, Richmond T, Somerville SC, Manners JM (2000) Coordinated plant defense responses in Arabidopsis revealed by microarray analysis. Proc Natl Acad Sci USA 97:11655–11660

Schöffl F, Raschke E, Nagao RT (1984) The DNA sequence analysis of soybean heat-shock factors and identification of possible regulatory promoter elements. EMBO J 3:2491–2497

Schnitzler J-P, Jungblut TM, Heller W, Köfferlein M, Hutzler P, Heinzmann U, Schmelzer E, Ernst D, Langebartels C, Sandermann H (1996) Tissue localization of u.v.-B-screening pigments and of chalcone synthase mRNA in needles of Scots pine seedlings. New Phytol 132:247–258

Schubert R, Fischer R, Hain R, Schreier PH, Bahnweg G, Ernst D, Sandermann H (1997) An ozone-responsive region of the grapevine resveratrol synthase promoter differs from the basal pathogen-responsive sequence. Plant Mol Biol 34:417–426

Schützendübel A, Polle A (2002) Plant responses to abiotic stresses: heavy metal-induced oxidative stress and protection by mycorrhization. J Exp Bot 53:1351–1365

Schützendübel A, Schwanz P, Teichmann T, Gross K, Langenfeld-Heyser R, Goldbold DL, Polle A (2001) Cadmium-induced changes in antioxidative systems, hydrogen peroxide content, and differentiation in Scots pine roots. Plant Physiol 127:887–898

Schulze-Lefert P, Dangl JL, Becker-André M, Hahlbrock K, Schulz W (1989) Inducible in vivo DNA footprints define sequences necessary for UV light activation of the parsley chalcone synthase gene. EMBO J 8:651–656

Seki H, Ichinose Y, Kato H, Shiraishi T, Yamada T (1996) Analysis of cis-regulatory elements involved in the activation of a member of chalcone synthase gene family (PsChs1) in pea. Plant Mol Biol 31:479–491

Seki M, Narusaka M, Abe H, Kasuga M, Yamaguchi-Shinozaki K, Carninci P, Hayashizaki Y, Shinozaki K (2001) Monitoring the expression pattern of 1300 Arabidopsis genes under drought and cold stresses by using a full-length cDNA microarray. Plant Cell 13:61–72

Sheen J (2002) Phosphorelay and transcription control in cytokinin signal transduction. Science 296:1650–1652

Shen Q, Ho T-HD (1995) Functional dissection of an abscisic acid (ABA)-inducible gene reveals two independent ABA-responsive complexes each containing a G box and a novel cis-acting element. Plant Cell 7:295–307

Shen Q, Zhang P, Ho T-HD (1996) Modular nature of abscisic acid (ABA) response complexes: composite promoter units that are necessary and sufficient for ABA induction of gene expression in barley. Plant Cell 8:1107–1119

Shimizu T, Akada S, Senda M, Ishikawa R, Harada T, Niizeki M, Dube SK (1999) Enhanced expression and differential inducibility of soybean chalcone synthase genes by supplemental UV-B in dark-grown seedlings. Plant Mol Biol 39:785–795

Shinozaki K, Yamaguchi-Shinozaki K (2000) Molecular responses to dehydration and low temperature: differences and cross-talk between two stress signaling pathways. Curr Opin Plant Biol 3:217–223

Stockinger EJ, Gilmour SJ, Thomashow MF (1997) Arabidopsis thaliana CBF1 encodes an AP2 domain-containing transcriptional activator that binds to the C-repeat/DRE, a cis-acting DNA regulatory element that stimulates transcription in response to low temperature and water deficit. Proc Natl Acad Sci USA 94:1035–1040

Suzuki N, Koizumi N, Sano H (2001) Screening of cadmium-responsive genes in *Arabidopsis thaliana*. Plant Cell Environ 24:1177–1188

Thomashow MF (1999) Plant cold acclimation: Freezing tolerance genes and regulatory mechanisms. Annu Rev Plant Physiol Plant Mol Biol 50:571–599

Thomma BPHJ, Eggermont K, Penninckx IAMA, Mauch-Mani B, Vogelsang R, Cammue BPA, Broekaert WF (1998) Separate jasmonate-dependent and salicylate-dependent defense-response pathways in *Arabidopsis* are essential for resistance to distinct microbial pathogens. Proc Natl Acad Sci USA 95:15107–15111

Tiwari SB, Wang X-J, Hagen G, Guilfoyle TJ (2001) AUX/IAA proteins are active repressors, and their stability and activity are modulated by auxin. Plant Cell 13:2809–2822

Ulmasov T, Ohimiya A, Hagen G, Guilfoyle T (1995) The soybean *GH2/4* gene that encodes a glutathione *S*-transferase has a promoter that is activated by a wide range of chemical agents. Plant Physiol 108:919–927

Ulmasov T, Hagen G, Guilfoyle TJ (1997) ARF1, a transcription factor that binds to auxin response elements. Science 276:1865–1868

Vernooij B, Uknes S, Ward E, Ryals J (1994) Salicylic acid as a signal molecule in plant-pathogen interactions. Curr Opin Cell Biol 6:275–279

Weisshaar B, Jenkins GI (1998) Phenylpropanoid biosynthesis and its regulation. Curr Opin Plant Biol 1:251–257

Werner T (2002) Finding and decrypting of promoters contributes to the elucidation of gene function. Electronic Publication, In Silico Biol. 2:0023 http://www.bioinfo.de/isb/2002/02/0023/main.html

Xiang C, Oliver DJ (1998) Glutathione metabolic genes coordinately respond to heavy metals and jasmonic acid in *Arabidopsis*. Plant Cell 10:1539–1550

Yamaguchi-Shinozaki K, Shinozaki K (1994) A novel *cis*-acting element in an *Arabidopsis* gene is involved in responsiveness to drought, low-temperature, or high-salt stress. Plant Cell 6:251–264

Yamaguchi-Shinozaki K, Mundy J, Chua N-H (1989) Four tightly linked rab genes are differentially expressed in rice. Plant Mol Biol 14:29–39

Yamamoto S, Suzuki K, Shinshi H (1999) Elicitor-responsive, ethylene-independent activation of GCC box-mediated transcription that is regulated by both protein phosphorylation and dephosphorylation in cultured tobacco cells. Plant J 20:571–579

Yamamoto Y, Kobayashi Y, Devi SR, Rikiishi S, Matsumoto H (2002) Aluminium toxicity is associated with mitochondrial dysfunction and the production of reactive oxygen species in plant cells. Plant Physiol 128:63–72

Yang P, Chen C, Wang Z, Fan B, Chen Z (1999) A pathogen- and salicylic acid-induced WRKY DNA-binding activity recognizes the elicitor response element of the tobacco class I chitinase gene promoter. Plant J 18:141–149

Yang Q, Grimmig B, Matern U (1998) Anthranilate *N*-hydroxycinnamoyl/ benzoyltransferase gene from carnation: rapid elicitation of transcription and promoter analysis. Plant Mol Biol 38:1201–1214

Yang Y, Shah J, Klessig DF (1997) Signal perception and transduction in plant defense responses. Genes Dev 11:1621–1639

Zhang B, Singh KB (1994) *ocs* element promoter sequences are activated by auxin and salicylic acid in *Arabidopsis*. Proc Natl Acad Sci USA 91:2507–2511

Zhang B, Chen W, Foley RC, Büttner M, Singh K (1995) Interactions between distinct types of DNA binding proteins enhance binding to *ocs* element promoter sequences. Plant Cell 7:2241–2252

Zhang B, Egli D, Georgiev O, Schaffner W (2001a) The *Drosophila* homolog of mammalian zinc finger factor MTF-1 activates transcription in response to heavy metals. Mol Cell Biol 21:4505–4514

Zhang Y, Chai T-Y, Dong J, Zhao W, An C-C, Chen Z-L, Burkhard G (2001b) Cloning and
    expression analysis of the heavy-metal responsive gene PvSR2 from bean. Plant Sci
    161:783–790
Zhou D-X (1999) Regulatory mechanism of plant gene transcription by GT elements and
    GT factors. Trends Plant Sci 4:210–214
Zhu J-K (2001) Cell signaling under salt, water and cold stresses. Curr Opin Plant Biol
    4:401–406

7 Increased Plant Fitness by Rhizobacteria

Leendert C. van Loon and Bernard R. Glick

7.1 Introduction

About half of the recently fixed carbon in plant leaves is transported below ground where a substantial fraction is released by growing plant roots as exudates and lysates. These nutrients attract bacteria and fungi, which multiply in the rhizosphere to densities up to and exceeding 100 times those in the bulk soil (Lynch and Whipps 1991). Some of these microorganisms can reduce plant growth by acting as pathogens. However, other microorganisms can promote growth by alleviating growth-restricting conditions (Schippers et al. 1987; Glick et al. 1999). Plant growth-promoting rhizobacteria (PGPR) can affect plant growth and development in two different ways: indirectly or directly (Glick 1995; Glick et al. 1999). Indirect promotion of plant growth occurs when these bacteria decrease or prevent some of the deleterious effects of a pathogenic organism by any one or more of several different mechanisms. For example, production of antibiotics can interfere directly with growth and activity of deleterious soil microorganisms (Glick and Bashan 1997), whereas induction of resistance in the plant increases the plant's defensive capacity (Van Loon et al. 1998). In addition, bacteria may reduce stresses resulting from the presence of toxic wastes by sequestering heavy metals or degrading organic pollutants. Direct promotion of plant growth by plant growth-promoting bacteria generally involves providing the plant with a compound that is synthesized by the bacterium or facilitating the uptake of nutrients.

There are several ways in which plant growth-promoting bacteria can directly facilitate the proliferation of their plant hosts. They may: fix atmospheric nitrogen; solubilize minerals such as phosphorus; produce siderophores, which can solubilize and sequester iron, and provide it to plants; synthesize phytohormones, including auxins, cytokinins, and gibberellins, which can enhance various stages of plant growth; and synthesize enzymes that can modulate plant growth and development (Brown 1974; Davison 1988; Kloepper et al. 1989; Lambert and Joos 1989; Patten and Glick 1996; Glick et al. 1999). In this way, PGPR can make a substantial contribution

Ecological Studies, Vol. 170
H. Sandermann (Ed.)
Molecular Ecotoxicology of Plants
© Springer-Verlag Berlin Heidelberg 2004

to plant growth and crop yield on poor soils. A particular bacterium may affect plant growth and development using any one, or more, of these mechanisms. Moreover, since many plant growth-promoting bacteria possess several traits that enable them to facilitate plant growth, a bacterium may utilize different traits at various times during the life cycle of the plant.

The mechanism most often invoked to explain the various effects of plant growth-promoting bacteria on plants is the production of phytohormones, most notably auxin (Brown 1974). Since plants as well as plant growth-promoting bacteria can synthesize indoleacetic acid (IAA), it is important when assessing the consequences of treating a plant with a plant growth-promoting bacterium to distinguish between the bacterial stimulation of plant auxin synthesis on the one hand and auxin that is synthesized by the bacterium on the other (Gaudin et al. 1994). The level of auxin produced by a bacterium in the rhizosphere determines its effect on the host plant: high levels induce developmental abnormalities and stimulate formation of lateral and adventitious roots, while low levels promote root elongation. To complicate matters further, the response of plants to auxin-producing bacteria may vary from one species of plant to another, as well as according to the age of the plant.

7.2 Rhizobacterially Induced Growth Promotion

7.2.1 Ethylene

Ethylene, which is produced in almost all plants, mediates a range of plant responses and developmental steps. Ethylene is involved in seed germination, tissue differentiation, formation of root and shoot primordia, root elongation, lateral bud development, flowering initiation, anthocyanin synthesis, flower opening and senescence, fruit ripening and degreening, production of volatile organic compounds responsible for aroma formation in fruits, storage product hydrolysis, leaf and fruit abscission and the response of plants to biotic and abiotic stresses (Mattoo and Suttle 1991; Abeles et al. 1992; Frankenberger and Arshad 1995). In some instances ethylene is stimulatory, while in others it is inhibitory.

The term "stress ethylene" was coined by Abeles (1973) to describe the increase in ethylene biosynthesis associated with biological and environmental stresses and pathogen attack (Morgan and Drew 1997). The increased level of ethylene formed in response to trauma inflicted by temperature extremes, water stress, ultraviolet light, chemicals, mechanical wounding, insect damage and disease can be both the cause of some of the symptoms of stress (e.g., onset of epinastic curvature and formation of aerenchyma) and the inducer of responses which will enhance survival of the plant under adverse conditions (e.g., cell wall strengthening, production of phytoalexins and synthesis of defensive proteins; Van Loon 1984).

Chemicals have been used to control ethylene levels in plants. The application of compounds such as rhizobitoxin, a non-protein amino acid secreted by some strains of *Rhizobia*, and its synthetic analog, aminoethoxyvinylglycine (AVG), can inhibit ethylene biosynthesis; silver thiosulfate can inhibit ethylene action, and 2-chloroethylphosphonic acid (ethephon), regarded by some researchers as "liquid ethylene," can release ethylene (Abeles et al. 1992). Sisler and Serek (1997) discovered that cyclopropenes can block ethylene perception and are potentially useful for extending the vase life of cut flowers and the display life of potted plants. All of these chemicals are potentially harmful to the environment: AVG and silver thiosulfate are highly toxic in food, and silver thiosulfate causes black spotting in flowers.

In higher plants ethylene is produced from L-methionine via the intermediates, S-adenosyl-L-methionine (SAM) and 1-aminocyclopropane-1-carboxylic acid (ACC) (Yang and Hoffman 1984). The enzymes involved in this metabolic sequence are SAM synthetase, which catalyzes the conversion of methionine to SAM (Giovanelli et al. 1980), ACC synthase, which is responsible for the production of ACC and 5′-methylthioadenosine from SAM (Kende 1989), and ACC oxidase, which further metabolizes ACC to ethylene, carbon dioxide and cyanide (John 1991).

7.2.2 ACC Deaminase

In 1978, an enzyme capable of degrading ACC was isolated from *Pseudomonas* sp. strain ACP and from the yeast, *Hansenula saturnus* (Honma and Shimomura 1978; Minami et al. 1998). Since then, ACC deaminase has been detected in the fungus, *Penicillium citrinum* (Honma 1993), and in a number of other bacterial strains (Klee and Kishore 1992; Jacobson et al. 1994, Glick et al. 1995; Campbell and Thomson 1996), all of which originated from the soil. Many of these microorganisms were identified by their ability to grow on minimal media containing ACC as its sole nitrogen source (Honma and Shimomura 1978; Klee et al. 1991; Honma 1993; Jacobson et al. 1994; Glick et al. 1995; Campbell and Thomson 1996; Burd et al. 1998; Belimov et al. 2001).

Enzymatic activity of ACC deaminase is assayed by monitoring the production of either ammonia or α-ketobutyrate, the products of ACC hydrolysis (Honma and Shimomura 1978). While ACC deaminase activity has been found only in microorganisms, there are no microorganisms that synthesize ethylene via ACC (Fukuda et al. 1993). In addition, throughout the many years that plants and microorganisms have been associated with one another, some plants may have acquired microbial ACC deaminase genes. Moreover, some of these genes may be expressed in the plant as a normal part of the plant genome (Jacobson et al. 1994). However, at the present time, there are no reports of ACC deaminase activity occurring naturally in plants.

ACC deaminase has been purified to homogeneity from *Pseudomonas* sp. strain ACP (Honma and Shimomura 1978) and partially purified from *Pseudomonas* sp. strain 6G5 (Klee et al. 1991) and *P. putida* GR12–2 (Jacobson et al. 1994); enzyme activity is localized exclusively in the cytoplasm (Jacobson et al. 1994). The molecular mass and form are similar for the ACC deaminase purified from all three sources. The enzyme is a trimer (Honma 1985); the size of the holoenzyme is approximately 104 to 105 kDa (Honma and Shimomura 1978; Honma 1985; Jacobson et al. 1994), and the subunit mass is approximately 36,500 Da (Honma and Shimomura 1978; Jacobson et al. 1994). Similar subunit sizes were predicted from nucleotide sequences of cloned ACC deaminase genes from *Pseudomonas* strains ACP (Sheehy et al. 1991) and 6G5 (Klee et al. 1991) and from *Enterobacter cloacae* UW4 (Shah et al. 1997).

K_m values for the binding of ACC by ACC deaminase have been estimated for enzyme extracts of 12 microorganisms at pH 8.5. These values ranged from 1.5 to 17.4 mM (Honma and Shimomura 1978; Klee and Kishore 1992; Honma 1993), indicating that the enzyme does not have a particularly high affinity for ACC (Glick et al. 1998).

ACC deaminase activity has been induced in both *Pseudomonas* sp. strain ACP and *P. putida* GR12–2 by ACC, at levels as low as 100 nM (Honma and Shimomura 1978; Jacobson et al. 1994); both bacterial strains were grown on a rich medium and then switched to a minimal medium containing ACC as its sole nitrogen source. The rate of induction, similar for the enzyme from the two bacterial sources, was relatively slow: complete induction required 8 to 10 h. Enzyme activity increased only approximately 10-fold over the basal level of activity, even when the concentration of ACC increased up to 10,000-fold.

Pyridoxal phosphate is a tightly bound cofactor of ACC deaminase in the amount of approximately 3 mol of enzyme-bound pyridoxal phosphate per mole of enzyme, or 1 mol per subunit (Honma 1985).

Genes encoding ACC deaminase have been cloned from a number of different soil bacteria, including *Pseudomonas* sp. strains 6G5 and 3F2 (Klee et al. 1991; Klee and Kishore 1992), *Pseudomonas* sp. strain 17 (Campbell and Thomson 1996), *Pseudomonas* sp. strain ACP (Sheehy et al. 1991) and *Enterobacter cloacae* strains CAL2 and UW4 (Glick et al. 1995; Shah et al. 1998), as well as from yeast, *Hansenula saturnus* (Minami et al. 1998), and fungi, *Penicillium citrinum* (Jia et al. 1999). In all of these instances the gene that was characterized was directly shown to encode a functional ACC deaminase. Putative ACC deaminase genes have been identified by DNA sequence homology in *Escherichia coli*, various Archaebacteria and the plant *Arabidopsis thaliana*; however, none of these organisms has been demonstrated to have ACC deaminase activity, and these genes cluster separately from the rhizobacterial and fungal ones (Fig. 7.1). In our laboratory, we have repeatedly assayed various strains of *E. coli* for ACC deaminase activity and have never

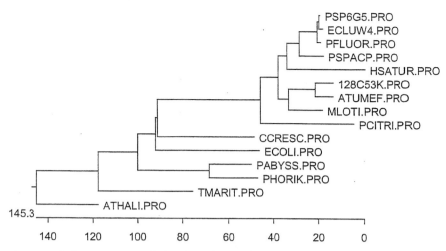

Fig. 7.1. Phylogenetic tree of the aligned ACC deaminase amino acid sequences. The length of each pair of branches represents the distance between sequence pairs. The *scale* beneath the *tree* measures the distance between the sequences, and the units indicate the number of substitution events. The *tree* was generated in MEGALIGN (DNAS-TAR Inc., Madison, Wis., USA) using Clustal method with PAM250 residue weight table. *PFLUOR Pseudomonas fluorescens*; *PSPACP Pseudomonas* spp. ACP; *PSP6G5 Pseudomonas* spp. 6G5; *ECLUW4 Enterobacter cloacae* UW4; *128C53K Rhizobium leguminosarum* bv. *viciae* 128 C53 K; *ATUMEF Agrobacterium tumefaciens*; *MLOTI Mesorhizobium loti*; *ECOLI Escherichia coli*; *CCRESC Caulobacter crescentus*; *PABYSS Pyrococcus abyssi*; *PHORIK Pyrococcus horikoshii*; *TMARIT Thermotoga maritima*; *PCITRI Penicillium citrinum*; *HSATUR Hansenula saturnus*; *ATHALI Arabidopsis thaliana*. The genes from *E. coli*, *C. crescentus*, *P. abyssi*, *P. horikoshii*, *T. maritima* and *A. thaliana* are all putative ACC deaminase genes based on DNA sequence analysis

detected the presence of this enzyme. It is likely that as the genomes of additional organisms are sequenced, many more putative ACC deaminase genes will be discovered. Nevertheless, it is essential that criteria in addition to sequence similarity be employed before it is decided that a gene actually encodes ACC deaminase rather than another similar protein.

The ACC deaminase genes from *Pseudomonas* sp. strains 6G5 and F17 and *E. cloacae* strains UW4 and CAL2 all have an ORF of 1,014 nucleotides that encode a protein containing 338 amino acids with a calculated molecular weight of approximately 36.8 kDa (Klee et al. 1991; Campbell and Thomson 1996; Shah et al. 1998). The genes from these strains are highly homologous to each other: at the nucleotide level 6G5, F17, UW4 and CAL2 are 85 to 95 % identical to each other (Campbell and Thomson 1996; Shah et al. 1998), and most of the dissimilarities are in the wobble position (Shah et al. 1998). However, the DNA sequences from strains UW4 and CAL2 show only about 74 % homology with the sequence of the ACC deaminase gene from *Pseudomonas* sp. strain ACP (Sheehy et al. 1991; Shah et al. 1998).

Sequence data indicate that upstream of the ACC deaminase gene from *E. cloacae* UW4 there is a DNA region that contains a CRP (cyclic AMP receptor protein) binding site, an FNR (fumarate-nitrate reduction regulatory protein) binding site (which is a known anaerobic transcriptional regulator), an LRP (leucine-responsive regulatory protein) binding site and an open reading frame encoding an LRP-like protein (Grichko and Glick 2000; Li and Glick 2001). It is thought that all of these features are involved in the transcriptional regulation of the ACC deaminase gene. Moreover, the *E. cloacae* UW4 ACC deaminase gene promoter appears to be under the transcriptional control of the LRP-like protein bound to ACC (Grichko and Glick 2000; Li and Glick 2001). This complex mode of regulation helps to ensure the appropriate levels of expression of ACC deaminase in a soil bacterium that exists at the aerobic-anaerobic interface (Fig. 7.2).

When a broad host range plasmid containing the ACC deaminase gene from *E. cloacae* UW4 was introduced into two non-plant growth-promoting bacteria, *P. putida* ATCC 17399 and *P. fluorescens* ATCC 17400, or into *Azospirillum brasilense*, by conjugational transfer, the transconjugants acquired the ability to grow on minimal media using ACC as the sole source of nitrogen and to promote the elongation of canola roots (Shah et al. 1998; Holguin and Glick 2001). For the ACC deaminase gene to be expressed in *A. brasilense*, it

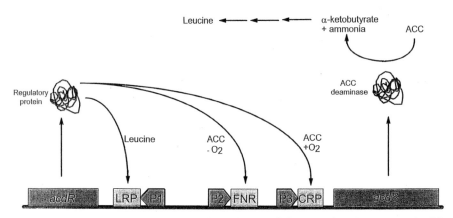

Fig. 7.2. A model of the transcriptional regulation of the ACC deaminase gene (*acdS*) from *Enterobacter cloacae* UW4. It is envisioned that the constitutive promoter P1 continuously directs the transcription of *acdR* resulting in the production of an LRP-like regulatory protein. If there is an excess of free leucine in the cell, the regulatory protein-leucine complex binds to the LRP box or to a transcription factor that binds to this box, thereby preventing any additional transcription of *acdR*. If there is ACC present, the regulatory protein-ACC complex binds to either the FNR box (under anaerobic conditions) or the CRP box (under aerobic conditions) or to a transcription factor(s) that binds to one or more of these boxes, thereby activating transcription of *acdS* from either promoter P2 or P3, respectively. The α-ketobutyrate that is formed from the cleavage of ACC is a metabolic precursor of leucine

was necessary to first replace the endogenous *E. cloacae* UW4 promoter region with another bacterial promoter such as the promoter from the *E. coli lac* operon (Holguin and Glick 2001).

7.2.3 A Model of Bacterial Lowering of Plant Ethylene Levels

A model has been proposed (Glick et al. 1998) in which plant growth-promoting bacteria can lower plant ethylene levels and in turn stimulate plant growth. In this model, the plant growth-promoting bacteria bind to the surface of either the seed or root of a developing plant. In response to tryptophan and/or other small molecules in the seed or root exudates (Whipps 1990), the plant growth-promoting bacteria synthesize and secrete IAA, some of which is taken up by the plant. This IAA, together with endogenous plant IAA, can stimulate plant cell proliferation and elongation or induce the activity of ACC synthase to convert SAM to ACC (Kende 1993).

Much of the ACC produced by this latter reaction is exuded from seeds or plant roots along with other small molecules normally present in seed or root exudates. Imbibed canola seeds exuded about 10 fmol ACC per seed per h (Penrose and Glick 2001), whereas root systems of 55-day-old tomato plants released micromolar concentrations of ACC upon flooding stress (Grichko and Glick 2001). The ACC in the exudates may be taken up by ACC deaminase-containing bacteria and subsequently cleaved by the enzyme to ammonia and α-ketobutyrate, thereby decreasing the amount of ACC outside the plant. Subsequently, increasing amounts of ACC are exuded by the plant in order to maintain the equilibrium between internal and external ACC levels (Glick et al. 1998). The net result of these interactions is that the bacteria induce the plant to synthesize more ACC than it would otherwise, and as well stimulate the exudation of ACC from the plant.

Thus, plant growth-promoting bacteria are supplied with a unique additional source of nitrogen in the form of ACC that enables them to proliferate under conditions in which other soil bacteria may not flourish, for instance, when nitrogen availability is low and competition for nutrients is intense. As a result of lowering the ACC level within the plant, either the endogenous level or the IAA-stimulated level, the amount of ethylene in the plant is also reduced.

Plant growth-promoting bacteria that possess the enzyme ACC deaminase and are bound to seeds or roots of seedlings can reduce the amount of plant ethylene and the extent of its inhibition on root elongation. Thus, these plants should have longer roots and possibly longer shoots as well, inasmuch as stem elongation is also inhibited by ethylene, except in ethylene-resistant plants.

Consistent with the model, when the ACC deaminase gene (*acdS*) from *Enterobacter cloacae* UW4 was replaced by homologous recombination with a version of the same gene with a tetracycline resistance gene inserted within the coding region, ACC deaminase activity was completely lost and the ability to promote the elongation of canola roots under gnotobiotic conditions was greatly diminished (Li et al. 2000). It was previously reported that ACC deaminase mutants of *Pseudomonas putida* GR12-2 did not promote the elongation of canola roots (Glick et al. 1994). However, in those experiments, the mutants were created by chemical mutagenesis, and as a result, one could never be certain that the mutations were within the ACC deaminase structural gene *per se*. In the experiments mentioned above (Li et al. 2000), ACC deaminase function was specifically eliminated by replacing the functional gene with an inactive version so that there is no longer any ambiguity as to the nature of the ACC deaminase minus mutants.

7.2.4 Treatment of Plants with ACC Deaminase-Containing Bacteria

Plants respond to a variety of different stresses including flooding, heavy metals and fungal infection by synthesizing "stress" ethylene (Van Loon 1984; Hyodo 1991). In turn, the ethylene can trigger a stress/senescence response in the plant, which may lead to the death of those cells that are at or near the site of the fungal infection. Stress ethylene is thought to act as a secondary messenger and can (in different tissues in different plants) stimulate senescence, leaf or fruit abscission, disease development, inhibition of growth and/or synthesis of defense-related enzymes (e.g., chitinase, glucanase and peroxidase).

Many of a plant's symptoms from an environmental stress occur as a result of the response of the plant to the increased levels of stress ethylene. And, not only does exogenous ethylene often increase the severity of the response, but, as well, inhibitors of ethylene synthesis can significantly decrease the severity of an environmental stress (Hall et al. 1996). Moreover, by acting as a sink for ACC, ACC deaminase-containing plant growth-promoting bacteria can act to modulate the level of ethylene in a plant.

7.2.4.1 Flower Senescence

Ethylene is a key signal in the initiation of flower wilting in most plants. Typically, flowers produce minute amounts of ethylene until an endogenous rise of the phytohormone, which is responsible for flower senescence (Mol et al. 1995), occurs. However, ethylene does not cause senescence in all flower families, and even the senescence symptoms that are caused by ethylene differ from plant to plant (Woltering and Van Doorn 1988). For example, Caryophyllaceae (e.g., carnations) show ethylene-mediated wilting of their petals, whereas ethylene causes petal abscission in Rosaceae (e.g., roses), and in Compositae (e.g., sunflowers) ethylene does not cause any senescence of petals.

Many cut flowers (e.g., carnations and lilies) are routinely treated with the ethylene inhibitor, silver thiosulfate, prior to their sale. However, high silver thiosulfate concentration is potentially phytotoxic and environmentally hazardous. On the other hand, use of ACC deaminase-containing plant growth-promoting bacteria to lower ACC levels in cut flowers might be an environmentally friendly alternative to silver thiosulfate. As a first step toward determining whether this suggestion is at all feasible, it was shown that the senescence of carnation petals that were treated with ACC deaminase-containing plant growth-promoting bacteria could be delayed by several days compared to untreated flower petals (Nayani et al. 1998).

7.2.4.2 Flooding

Flooding is a common biotic stress that is faced by many plants, often several times during the same growing season. A lack of oxygen in the roots is the main consequence of flooding; this in turn causes deleterious symptoms such as epinasty, leaf chlorosis, necrosis and reduced fruit yield. In flooded tomato plants two of the ACC synthase genes are rapidly induced in the roots. Since ACC oxidase-catalyzed ethylene synthesis can not occur in the anaerobic environment of flooded roots, the newly synthesized ACC is transported into the aerobic shoots where it is converted to ethylene, resulting in the deleterious effects indicated above (Bradford and Yang 1980; Else and Jackson 1998). In addition, as a part of their response to the flooding stress, plants exude ACC into the surrounding water, which may stimulate the proliferation of facultative anaerobic bacteria with ACC deaminase activity (Grichko and Glick 2001). Treatment of tomato plants with ACC deaminase-containing plant growth-promoting bacteria significantly decreases the damage to these plants that would normally occur as a consequence of flooding (Grichko and Glick 2001). Protection against flooding stress, which was achieved as a result of the presence on the roots of treated tomato plants of ACC deaminase-containing bacteria, resulted in statistically significant differences in overall plant growth, leaf chlorophyll content and substantially decreased ethylene production in leaf petiolar tissue (Grichko and Glick 2001). Thus, the "protective effect" of ACC deaminase-containing plant growth-promoting bacteria on flooded tomato plants results from these bacteria acting as a sink for ACC, thereby lowering the level of ethylene that can be formed in the shoots.

7.2.4.3 Metal Stress

Reasoning that a portion of the inhibitory effect of metals on plant growth is likely the consequence of the plant synthesizing an excessive amount of stress ethylene in response to the presence of the metal, especially during early seedling development, a metal-resistant ACC deaminase-containing bacterium was added to canola and tomato seeds prior to planting these seeds in metal-contaminated soil. Seeds inoculated with this bacterium, *Kluyvera ascorbata*, and then grown in the presence of high concentrations of nickel chloride were partially protected against nickel toxicity (Burd et al. 1998). The presence of this bacterium had no measurable influence on the amount of nickel accumulated per mg dry weight of either roots or shoots of canola plants. Therefore, the bacterial plant growth-promoting effect in the presence of nickel was not attributable to a reduction of nickel uptake by seedlings. Rather, it reflects the ability of the bacterium to lower the level of stress ethylene caused by the nickel.

7.2.4.4 *Rhizobia* Infection

Considerable evidence suggests that the ethylene that is produced following infection of legumes by strains of *Rhizobia* is inhibitory to the process of nodulation. For example, an ethylene-insensitive mutant of alfalfa that can be hyperinfected by its symbiotic partner has been reported (Penmetsa and Cook 1997). In addition, there is a report of an ethylene-hypersensitive mutant of pea plants, which forms fewer nodules compared with the wild-type plants (Guinel and Sloetjes 2000).

In preliminary experiments, it was observed that five of twelve rhizobial strains that were tested contained the enzyme ACC deaminase (Ma et al. 2003). Moreover, inactivation of this gene significantly decreased the number of nodules that the bacterium formed with pea plants.

7.2.4.5 Pathogen Infection

Ethylene can have a profound effect on the severity of a fungal infection. For example: (1) in a study of the reaction of over 60 different cultivars and breeding lines of wheat to the fungal pathogen *Septoria nodorum*, increased ethylene production as a result of the fungal infection was correlated with increased plant disease susceptibility (Hyodo 1991); (2) treatment of cotton plants with chemical inhibitors of ethylene synthesis decreased leaf blight caused by the fungus *Alternaria* (Bashan 1994); (3) melon plants that were treated with inhibitors of ethylene synthesis showed decreased levels of ethylene and decreased disease severity following infection by the fungus *Fusarium oxysporum* (Cohen et al. 1986); (4) pretreating plants with ethylene increased fungal disease development (a) in cucumber caused by *Colletotrichum lagenarium* (Biles et al. 1990) and (b) in tomato caused by *Verticillium dahliae* (Cronshaw and Pegg 1976). Treatment of roses, carnations, tomato, pepper, French-bean and cucumber with ethylene inhibitors decreased gray mold caused by the fungus *Botrytis cinerea* (Elad 1988, 1990).

In one set of experiments, the effect of transforming several PGPR strains with the *Enterobacter cloacae* UW4 ACC deaminase gene was assessed based on the suppression of damping off of cucumber caused by *Pythium ultimum* (Wang et al. 2000). These experiments indicated that ACC deaminase-containing bacterial strains were significantly more effective than biocontrol strains that did not possess this enzyme. Moreover, transgenic tomato plants that express ACC deaminase are also protected to a significant extent against damage from several different pathogens (Lund et al. 1998; Robison 2001). In effect, ACC deaminase acts synergistically with other mechanisms of biocontrol to prevent pathogens from damaging plants. As with other types of stress, it is assumed that ACC deaminase can act to prevent ACC build up that would otherwise occur as a result of environmental stress.

7.3 Rhizobacteria-Mediated Induced Systemic Resistance

7.3.1 Disease Suppression by Plant Growth-Promoting Bacteria

Besides promoting growth and alleviating stress symptoms, several rhizobacterial strains can protect plants directly against harmful microorganisms including fungi, bacteria, viruses as well as nematodes (Fig. 7.3). The mechanisms involved are multiple and comprise both a direct action of the rhizobacteria against the attacker and a stimulation of defensive mechanisms in the plant (Handelsman and Stabb 1996; Van Loon et al. 1998). Bacterial antagonism can take the form of niche exclusion at the root surface, competition for nutrients required for spore germination, fungal or bacterial growth or pathogenic activity, production of antimicrobial compounds that affect fungi or bacteria and secretion of lytic enzymes, such as chitinases, glucanases and proteases, that are effective in disrupting growing fungal hyphae or damage nematode structures. These mechanisms are effective only when the protective bacteria and the attacking organism are in close proximity. Hence, in the soil they are confined to the environment directly surrounding plant roots, with disease suppression being restricted to soil-borne pathogens. By contrast, rhizobacterial stimulation of plant defenses is mostly systemic and can protect plants not only against root pathogens but also against foliar pathogens. This phenomenon is commonly known as rhizobacteria-mediated systemic acquired resistance (SAR) or induced systemic resistance (ISR) (Kloepper et al. 1992; Hammerschmidt et al. 2001). However, different terminologies have been put forward to distinguish mechanistically different forms of systemically induced resistance (Pieterse et al. 1996; Van Loon 1997).

Fig. 7.3. (in color at end of book) Reduction of Fusarium wilt in radish by biocontrol bacteria under commercial greenhouse conditions; *left plot* seeds treated with a coating containing *Pseudomonas fluorescens* strain WCS374; *middle plot* coating without bacteria; *right plot* non-treated seeds

Systemic resistance induced by specific rhizobacterial strains was first described in 1991 by Van Peer et al. in carnation against *Fusarium oxysporum* f.sp. *lycopersici* and by Wei et al. in cucumber against *Colletotrichum orbiculare*. In both cases the inducing rhizobacterial strain(s) remained confined to the roots, and the pathogen was inoculated on, and colonized only, the above-ground plant parts. No contact occurred between the inducing rhizobacterium and the challenging pathogen. Thus, the resulting disease suppression must be a plant-mediated phenomenon. For carnation it was shown that a crude cell wall preparation from the inducing bacterium was as active in inducing systemic resistance as live bacteria (Van Peer and Schippers 1992), ruling out any direct effect of a transported bacterial metabolite on the attacking fungus. In the last decade, rhizobacteria-mediated ISR has likewise been shown to occur in bean, radish, tobacco, tomato, as well as in the model plant *Arabidopsis thaliana*, and to be effective against pathogenic fungi, bacteria and viruses (Maurhofer et al. 1994; Alström 1995; Leeman et al. 1995a; Pieterse et al. 1996; Duijff et al. 1998).

7.3.2 Bacterial Signals Involved in Induction of Systemic Resistance

How rhizobacteria trigger systemically induced resistance is not known, but the following sequence of events is envisaged. A bacterial component or

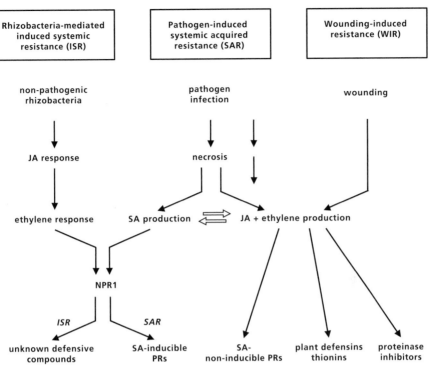

Fig. 7.4. Signalling pathways involved in rhizobacteria-mediated induced systemic resistance (*ISR*), pathogen-induced systemic acquired resistance (*SAR*) and wound-induced resistance (*WIR*). PRs, plant defensins and thionins comprise antifungal and antibacterial proteins; proteinase inhibitors are directed against insects and nematodes. *JA* Jasmonic acid; *NPR1* non-expressor of PRs; *PRs* pathogenesis-related proteins; *SA* salicylic acid. (After Pieterse and Van Loon 1999)

metabolite is perceived by the plant root through binding to a receptor. Thereupon, the metabolite itself or a signal generated by the plant cells as a result of ligand binding is transported throughout the entire plant, i.e., to both the leaves and to the other roots. It is possible that an amplification step is required in this process. Finally, the translocated signal is perceived by distant plant cells and a signal-transduction pathway is activated that gives rise to the establishment of the systemically enhanced defensive capacity. This sequence of events is similar to wound-induced resistance (WIR) in tomato leaves (Fig. 7.4), which gives rise to the accumulation, e.g., of proteinase inhibitors (PIs) as a defense against herbivorous insects (Farmer and Ryan 1992; Ryan 1992). Endogenous oligogalacturonide elicitors that are released upon insect feeding or cell damage, and also exogenous cell wall elicitors from infecting pathogens, bind to a receptor on the leaf cell plasma membrane. A phospholipase is activated, which leads to the release of linolenic acid, which is converted in the octadecanoid pathway to jasmonic acid (JA). Moreover, an 18-

amino-acid peptide, systemin, is generated by cleavage of the cellular protein prosystemin. Systemin is transported systemically and acts similarly to elicitors in binding to a receptor that functions in the signalling pathway leading to the production of JA. The JA then acts locally – but may also be transported systemically – as a hormone and induces proteinase inhibitors.

Rhizobacterially mediated systemically induced resistance is the result of a process that does not involve any obvious damage to the plant. The inducing bacteria are fully saprophytic and can induce resistance and promote plant growth at the same time (Tuzun and Kloepper 1995; Pieterse and Van Loon 1999). In a few cases, bacterial determinants that are responsible for the induction of resistance have been identified through transposon mutagenesis and application of the purified factors involved. In this way it was established that different bacterial strains that are able to elicit ISR in radish against *Fusarium oxysporum* f.sp. *raphani* use both similar and divergent factors. Of strains *Pseudomonas putida* WCS358, *Pseudomonas fluorescens* WCS374 and *P. fluorescens* WCS417, only the latter two induce systemic resistance in radish (Leeman et al. 1995a). All three strains have outer membrane lipopolysaccharide (LPS) with strain-specific O-antigenic carbohydrate side-chains. The latter act as inducing determinants of WCS374 and WCS417 (Leeman et al. 1995b). Furthermore, all three strains are fluorescent pseudomonads that produce strain-specific pyoverdin siderophores under iron-limiting conditions. The purified siderophore of WCS374 (pseudobactin 374) could also induce resistance in radish, whereas those of WCS358 and WCS417 could not (Leeman et al. 1996).

Moreover, (an) additional iron-regulated metabolite(s) of WCS374 and WCS417 was implicated in the elicitation of ISR, because mutants lacking the O-antigenic side chain of the LPS and/or the pseudobactin siderophore were still capable of inducing systemic resistance in radish under iron-limiting conditions (Leeman et al. 1996). This capacity was correlated with the ability of the strains to produce salicylic acid (SA) in vitro, where it may act as an additional siderophore when iron is scarce (Meyer et al. 1992). SA is a central regulator in pathogen-induced SAR (Ryals et al. 1996; Durner et al. 1997). SAR is another type of systemically induced resistance (Fig. 7.4), where limited pathogen infection – often in the form of a hypersensitive reaction – leads to a state that is phenotypically similar to rhizobacteria-mediated ISR in being effective against different types of pathogens. In *Arabidopsis*, tobacco and tomato plants transformed with the *NahG* gene from a *P. putida* strain, which encodes salicylate hydroxylase, any SA formed is converted into catechol, which is inactive in inducing SAR. These transformants cannot express SAR, demonstrating that SA is essential for the establishment of SAR. SAR is associated with the induction, by SA, of defense-related pathogenesis-related proteins (PRs), and these serve as convenient molecular markers of the state of SAR (Ryals et al. 1996; Van Loon 2000).

If resistance-inducing rhizobacteria produce SA, it could be expected that the SA is taken up by plant roots and could act directly as the signal for the induction of systemic resistance. However, at least in radish induced by strains WCS374 or WCS417, no induction of PRs was apparent (Hoffland et al. 1995). This suggests that in the induction of systemic resistance by these strains SA does not act as an inducing determinant in the rhizosphere. However, treatment of the roots of radish plants with only nanogram amounts of SA sufficed to induce systemic resistance against *F. oxysporum* f.sp. *raphani* (Leeman et al. 1996). Those concentrations of SA are too low to induce PRs. Therefore, it cannot be ruled out that small amounts of SA produced by the inducing bacterial strains are responsible for the induction of systemic resistance in radish. The biocontrol bacterium *Pseudomonas aeruginosa* strain 7NSK2 is also capable of producing SA and has been demonstrated to induce resistance in bean, tomato and tobacco against anthracnose and gray mold, caused by *Colletotrichum lindemuthianum* and *Botrytis cinerea*, respectively, and against tobacco mosaic virus (De Meyer and Höfte 1997; De Meyer et al. 1999; Höfte et al. 2000). A mutant incapable of producing SA no longer induced resistance in these species. However, SA is synthesized by this bacterium as part of a pyochelin siderophore and induction by pyochelin rather than SA in the wild-type strain is now the favored hypothesis.

7.3.3 Induced Resistance Signalling in the Plant

To test whether SA could be involved in the induction of systemic resistance in radish by the *P. fluorescens* strains WCS374 and WCS417, induction of resistance was studied in the *NahG* transformant and several mutants of the related cruciferous species *Arabidopsis thaliana*, accession Columbia (Col). Using the bacterial leaf pathogen, *Pseudomonas syringae* pv. *tomato* for challenge inoculation, WCS417 was found to trigger ISR in both wild-type and *NahG* plants. WCS374, which activated ISR in radish, did not in *Arabidopsis*, whereas WCS358, which did not trigger ISR in radish, was as effective in both wild-type and *NahG Arabidopsis* plants as was WCS417 (Van Wees et al. 1997). Commonly, symptoms of *P. syringae* pv. *tomato* consisting of chlorosis progressing into necrosis were restricted to about half of the leaves compared to challenged non-induced plants. Thus, ISR was expressed as a substantial reduction in bacterial speck symptoms in *Arabidopsis* in a bacterial strain-specific manner, largely different from that in radish. Because the inducing strains were as active in triggering ISR in the *NahG* transformant as in wild-type plants, SA is not required for the ISR activated in *Arabidopsis*. Moreover, no accumulation of the PR-1, -2 and -5 mRNAs or proteins was observed in induced plants, supporting the non-involvement of SA and indicating that *PR* gene expression is not necessary for ISR to be expressed. Thereupon, the *Arabidopsis* mutant *npr1*, which is affected in SA action and cannot mount SAR or express PRs when challenged with a pathogen, was tested. Surprisingly, neither WCS417 nor WCS358 were capable of eliciting ISR in *npr1* plants, indicating that a functional NPR1 protein is required for the expression of rhizobacterially mediated ISR (Fig. 7.4), as it is for pathogen-induced SAR (Pieterse et al. 1996).

For induction of ISR in radish 10^5 colony-forming units of the inducing bacterial strain per gram root constitute a minimum threshold amount (Raaijmakers et al. 1995). This level is easily reached and maintained on bacterized roots of plants grown in autoclaved soil, but does not occur commonly with individual strains in natural soils. In experiments with autoclaved soil the three strains tested all colonized *Arabidopsis* roots to similarly high levels (Pieterse et al. 1996), ruling out the possibility that non-induction of ISR by WCS374 was due to poor root colonization. Hence, *Arabidopsis* roots perceive bacterial inducing determinants differently from radish. Extrapolating from this observation, it may be assumed that for each plant species different specific interactions occur between bacterial factors and root cells. In the case of ISR triggered by WCS358 in *Arabidopsis* both the LPS and the pseudobactin siderophore appear capable of inducing resistance (Bakker et al., 2003), as does a crude cell wall preparation from WCS417 (Van Wees et al. 1997). How these factors are perceived at the root surface is not known. Recently, a receptor for bacterial flagellin was characterized from *Arabidopsis* roots that recognizes a conserved domain within the flagella protein (Gómez-Gómez and Boller 2000). Thus, *Arabidopsis* roots are capable of recognizing the presence of bacteria through their flagella. Binding of flagellin leads to a reduction in root elongation (Gómez-Gómez et al. 1999). In contrast, most resistance-inducing rhizobacteria promote growth and enhance root elongation (Glick et al. 1999), and the flagellin-induced growth inhibition must be compensated by other activities. Purified flagella of WCS358 are capable of eliciting ISR in *Arabidopsis* (Bakker et al. 2003), and the conserved flagellin domain involved in binding is likely to be present in most, if not all, rhizobacteria. However, the elicitation of ISR in *Arabidopsis* is clearly strain-specific and most likely occurs through other, non-conserved parts of the flagellin.

7.3.4 Involvement of Ethylene and Jasmonic Acid in Induced Systemic Resistance

Within the species *Arabidopsis thaliana* the majority of accessions are inducible by WCS417 or WCS358, but ISR is not expressed in accessions RLD and Ws-0 (Van Wees et al. 1997; Ton et al. 1999). Root colonization by the inducing bacteria occurs to the same levels as on the accession Col, indicating that the two non-responsive accessions must either be deficient in perception or lack a component for ISR to be expressed. In view of the redundancy of the bacterial inducing determinants, the first possibility appeared unlikely. Interestingly, both RLD and Ws-0 stand out as being highly susceptible to the challenging pathogen, *P. syringae* pv. *tomato*, with symptoms developing more quickly and being more severe than in Col, for example. Crosses between RLD and Ws-0, on the one hand, and of RLD and Ws-0 with the responsive accessions Col and Landsberg *erecta* (L*er*), on the other hand, established that RLD and Ws-0 shared a recessive allele of a single gene that in Col and L*er* conferred both a basal level of resistance to *P. syringae* pv. *tomato* and the ability to express ISR. The corresponding locus was designated *Isr1* (Ton et al. 1999). When tested for responsiveness to ethylene, RLD and Ws-0 proved less sensitive to ethylene-induced inhibition of primary root growth. Also, upon application of the ethylene precursor ACC, expression of the ethylene-responsive genes *Hel* (encoding a hevein-type chitinase) and *Pdf1.2* (plant defensin) was reduced (Ton et al. 2001). In subsequent crosses the reduced sensitivity to ethylene cosegregated with the recessive alleles of the *Isr1* locus, i.e., with the inability to express ISR. It was concluded that the *Isr1* locus encodes a component of the ethylene response that is required for the expression of rhizobacteria-mediated ISR (Ton et al. 2001).

The *Arabidopsis* ethylene-response mutants *etr1* and *ein2* through *ein7*, as well as *jar1* – which has reduced sensitivity to JA – are also affected in their expression of ISR. In contrast, their ability to express pathogen-induced, SA-dependent SAR is unimpaired (Pieterse et al. 1998; Knoester et al. 1999). Application of either ACC or JA to wild-type Col plants induces a resistance that, like ISR, is not associated with the accumulation of PRs, but is dependent on a functional *Npr1* gene. Treatment of the *jar1* mutant with ACC was effective in inducing resistance. In contrast, application of JA to the *etr1* mutant did not elicit ISR. Moreover, ACC and JA were both ineffective in inducing resistance in the *npr1* mutant. These results established that responsiveness to JA and ethylene are required sequentially, and before NPR1, in the ISR signal-transduction pathway (Pieterse et al. 1998; Fig. 7.4).

Surprisingly, both at the site of application of the resistance-inducing rhizobacteria and systemically in leaves to be challenge-inoculated with the pathogen, JA content and ethylene evolution remained unaltered (Pieterse et al. 2000). Also, transformed plants in which expression of the lipoxygenase 2 (*lox2*) gene was impaired and that, consequently, are blocked in the production of JA upon wounding or pathogen infection, were normally responsive to induction treatments, indicating that ISR can be expressed in the absence of increased JA levels. One explanation could be that the JA and ethylene dependency of ISR is not linked to enhanced JA and ethylene production, but rather involves increased sensitivity to these hormones. However, inducing bacteria do not sensitize the plant to express JA- and/or ethylene-inducible genes. Of several known defense-related genes, the ethylene-inducible gene *Hel*, the ethylene- and jasmonate-responsive chitinase B (*ChiB*) and *Pdf1.2* genes and the JA-inducible genes coding for the vegetative storage protein Atvsp, lipoxygenase 1 (*Lox1*), Lox2, phenylalanine ammonia-lyase 1 (*Pal1*) and proteinase inhibitor 2 (*Pin2*) are neither induced locally in the roots nor systemically in the leaves upon triggering of ISR by WCS417 (Van Wees et al. 1999). However, whereas challenge inoculation with virulent *P. syringae* pv. *tomato* of plants induced by inoculation of three leaves with avirulent *P. syringae* pv. *tomato* and expressing SAR showed enhanced expression of *PR-1* mRNA, challenge inoculation of plants growing in soil containing WCS417 and expressing ISR led to an enhanced level of *Atvsp* transcript accumulation (Van Wees et al. 2000). Other JA-responsive defense-related genes studied were not potentiated during ISR, indicating that ISR is associated with the potentiation of specific JA-responsive genes. Such potentiation – also known as "priming" or "sensitization" (Durner et al. 1997; Thulke and Conrath 1998) – is also evident as a higher level of PR gene transcript accumulation during the expression of pathogen-induced SAR.

7.3.5 Differential Expression of Induced Systemic Resistance Against Different Pathogens

In *Arabidopsis*, rhizobacteria-mediated ISR has been found to be effective against the fungal root pathogen *F. oxysporum*, the fungal leaf pathogen *Alternaria brassicicola* and the bacterial leaf pathogens *P. syringae* pv. *tomato* and *Xanthomonas campestris* pv. *armoraciae*, and to a lesser extent against the oomycetous leaf pathogen *Peronospora parasitica* (Ton et al. 2002a). ISR was not effective against turnip crinkle virus. The latter result contrasts with other data showing that in cucumber and tomato ISR activated by rhizobacteria or by plant growth-promoting fungi is effective against cucumber mosaic virus and tomato mottle virus, as it is against the fungi *C. orbiculare* and *F. oxysporum* f.sp. *cucumerinum* and the bacteria *Erwinia tracheiphila* and *Pseudomonas syringae* pv. *lachrymans* (Tuzun and Kloepper 1995; Koike et al. 2001; Zehnder et al. 2001). *P. aeruginosa* strain CHA0 has been shown to induce resistance in tobacco against tobacco necrosis virus, as well as against the root rotting fungus *Thielaviopsis basicola* (Maurhofer et al. 1994; Troxler et al. 1997), as does *P. aeruginosa* 7NSK2 in tobacco against TMV, as well as in bean and tomato against *Colletotrichum lindemuthianum*, *Botrytis cinerea* and *Oidium lycopersici* (Höfte et al. 2000). The resistance induced by *P. aeruginosa* is fully associated with the ability of the bacterium to produce SA, implicating SA-dependent defenses in resistance against viruses. Indeed, resistance against viruses has recently been suggested to depend on SA by a signalling pathway different from that leading to resistance against fungi and bacteria (Murphy et al. 1999). The lack of effectiveness of WCS417-induced ISR in *Arabidopsis* can be explained by its non-dependency on SA. Apparently, the range of pathogens that is resisted by the systemic resistance induced by specific rhizobacterial strains depends on the signalling pathway that is activated in the plant.

This notion has been reinforced by findings that in *Arabidopsis* biotrophic pathogens are resisted mostly through SA-dependent defenses and necrotrophic pathogens more strongly through JA/ethylene-dependent defenses (Thomma et al. 1998) (Fig. 7.5). *NahG Arabidopsis* plants show enhanced susceptibility to the biotrophic downy mildew *Peronospora parasitica*, but not to the necrotrophic fungi *A. brassicicola* or *B. cinerea*. Conversely, JA-insensitive *coi1* and ethylene-insensitive *ein2* plants are more sensitive than wild-type plants to *B. cinerea*, but not to *P. parasitica*, whereas *A. brassicicola* is a pathogen on mutant plants that are impaired in the production of the phytoalexin camalexin (Thomma et al. 2001a,b). We recently confirmed these findings and established that the bacterial leaf pathogens *P. syringae* pv. *tomato* and *Xanthomonas campestris* pv. *armoraciae* are resisted by both SA- and JA/ethylene-dependent defenses, as both *NahG* and *jar1* and *ein2* plants show more severe symptoms upon infection with either of these pathogens (Ton et al. 2002a,c).

Largely in accordance with these differential sensitivities, SA-dependent SAR is very effective against *P. parasitica* and turnip crinkle virus, whereas rhizobacteria-mediated ISR is only moderately so against *P. parasitica* and not at all against the virus. *A. brassicicola* is resisted only moderately by either SAR or ISR. SAR and ISR are equally effective against *X. campestris*, whereas against *P. syringae* SAR is slightly more effective than ISR. These results clearly suggest that induced resistance acts by the same pathways as primary resistance (Fig. 7.5). Moreover, induced resistance is expressed as a potentia-

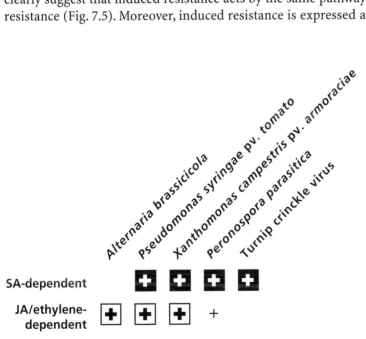

Fig. 7.5. *Arabidopsis* pathogens that are sensitive to the expression of SAR (*white plus signs*) are resisted through SA-dependent resistance (*black squares*), whereas pathogens sensitive to *Pseudomonas fluorescens* WCS417r-mediated ISR (*black plus signs*) are predominantly resisted through JA/ethylene-dependent resistance (*white squares*). (After Ton et al. 2002 c)

tion of extant resistance mechanisms. The latter can be further illustrated by the analysis of enhanced disease susceptibility (*eds*) mutants in *Arabidopsis*, which are more sensitive to specific pathogens than their wild types (Ton et al. 2002b). Several of these mutants are also impaired in the expression of either SAR or ISR because of blocks in the SA and JA/ethylene pathway, respectively.

7.3.6 Interactions Between Induced Resistance Signalling Pathways

Because SAR and ISR are both dependent on a functional NPR1 protein, it was investigated whether the SAR and the ISR pathways compete with each other or can act in an additive way in induced resistance against a pathogen such as *P. syringae*, which is resisted by both pathways. The resistance that was induced by both root colonization of WCS417 (ISR) and leaf infiltration of an avirulent strain of the pathogen *P. syringae* (SAR) was greater than that induced by either treatment alone. The enhancement by SAR was abolished in *NahG*, and by ISR in *jar1* and *etr1* plants, indicating that both signalling pathways contributed in an additive fashion (Van Wees et al. 2000). These results demonstrate that the level of induced resistance can be increased by simultaneous activation of both the ISR and the SAR pathway.

Whereas SAR and ISR can act in an additive way, antagonism is evident between SA-dependent SAR and JA-mediated wound-induced resistance against insects (Bostock 1999; Felton and Korth 2000; Paul et al. 2000). In transgenic plants with constitutively elevated levels of SA, SAR is also expressed constitutively. When such plants are attacked by larvae of the boll worm *Heliothis virescens*, the expression of JA-induced resistance is reduced. Conversely, in *NahG* plants, which cannot express SAR, JA-dependent induced resistance is enhanced. These effects result from cross-talk between the SA- and JA-dependent signalling pathways, such that activation of the former reduces JA-dependent responses and vice versa (Felton et al. 1999). This cross-talk is evident by an increased production of JA in wounded *NahG* plants, as well as an inability of SA to repress JA-inducible genes such as *Lox2* and *Pdf1.2* in SA-insensitive *npr1* plants (Spoel et al. 2003). A direct consequence of the enhanced resistance against pathogens afforded by SAR is, thus, increased vulnerability to attack by chewing insects. Interestingly, sucking insects activate, and appear to be affected by, SA-dependent defenses and are not more damaging on SAR-expressing plants (Walling 2000). The interaction between rhizobacteria-mediated ISR and insect-induced resistance has not been studied yet. Both these types of resistance depend on JA. However, wound-induced resistance involves an increase in the level of JA and the induction of JA-dependent defense-related genes, whereas in ISR no increase in JA occurs and defense-related genes are not induced. Moreover, wound-induced resistance does not depend on NPR1, indicating that the two pathways are separate and cross-talk is likely to be minimal. If so, additive effects between rhizobacteria-mediated ISR and wound-induced resistance may be expected, similar to what was shown for the combination of SAR and ISR.

The complexity of resistance signalling implies that distinct sets of responses are induced that enable the plant to adjust its defenses to the type of attacker encountered. Using microarrays, it is now becoming possible to analyze the transcriptome of plants and make a full inventory of the changes in gene expression in response to specific treatments. cDNA microarray analysis of *Arabidopsis* leaves inoculated with *A. brassicicola* or treated with SA, methyl-JA or ethylene has already shown that hundreds of genes are up- or down-regulated as a result of one or more of these treatments (Schenk et al. 2000). The changes indicate the existence of a substantial network of regulatory interactions and coordination, notably between the SA and JA signalling pathways. Substantial changes in gene expression were also described for *Arabidopsis* leaves in response to mechanical wounding and insect feeding (Reymond et al. 2000), as well as during chemically or pathogen-induced SAR and in constitutively SAR-expressing mutant plants (Maleck et al. 2000). In contrast, so far, few changes in gene expression are apparent in plants with ISR elicited by rhizobacteria (Van Wees et al. 2000).

7.4 Concluding Remarks

Several rhizobacterial strains actively promote plant growth and suppress diseases. Also, various soil-borne fungi, including mycorrhizae and endophytes, can stimulate plant growth and reduce diseases. Mycorrhizae extend the volume of soil that can be explored and aid the plant particularly in the uptake of phosphorus, whereas growth promotion by endophytic microorganisms is likely due to hormonal stimulation. Whether these fungi suppress disease mainly through competition with pathogens or also act by inducing resistance in the plant is not always clear, because it is difficult to demonstrate experimentally that the fungus and the pathogen do not interact directly. Yet, the ability of nonpathogenic rhizobacteria and fungi to induce systemic resistance appears to be fairly common, but by no means general (Van Loon and Pieterse 2002). Microbial activities exerted in the rhizosphere influence plant growth, development and metabolism at both the root and the shoot levels and can reduce the effects of various stresses. In general, rhizobacteria display little specificity in the plant species that they can colonize (Glandorf et al. 1993). Consequently, the roots of individual plants are colonized by many different microbial species and strains, and intense competition for nutrients limits their population densities. Introduction of specific rhizobacterial strains with desirable properties into the root environment requires high numbers of cells in order to shift the dynamic microbial balance in favor of the introduced strain. Any special trait that promotes competitive root colonization will aid in the establishment of the introduced bacterium (Naseby and Lynch 1998; Walsh et al. 2001). Several PGPR strains with different properties are already commercially available and are being used to increase crop yields, e.g., in Russia, China and Latin America. Given the current reluctance on the part of many consumers worldwide to embrace the use as foods of genetically modified plants, for the foreseeable future it may be advantageous to use either natural or genetically engineered plant growth-promoting bacteria as a means to promote growth by lowering plant ethylene levels or reduce disease through induction of resistance, rather than genetically modifying the plant itself to the same end (O'Connell et al. 1996). Moreover, given the large number of different plants, the various cultivars of those plants and the multiplicity of genes that would need to be engineered into plants, it is not feasible to genetically engineer all plants to be resistant to all pathogens and environmental stresses. Rather, it makes a lot more sense to engineer plant growth-promoting bacteria to do this job, and the first step in this direction could well be the introduction of appropriately regulated ACC deaminase genes. While ethylene signalling is required for the induction of systemic resistance elicited by rhizobacteria, a significant increase in the level of ethylene is not. Hence, lowering of ethylene levels by bacterial ACC deaminase does not appear to be incompatible with the induction of systemic resistance. Indeed, some bacterial strains possessing ACC deaminase also induce systemic resistance. What plant genes become expressed during these processes, what regulatory factors are involved and how these processes interact are questions that need to be further addressed in order to obtain a fuller understanding of how plants can cope with various stresses by benefitting from their rhizosphere microflora.

Acknowledgements. We thank the Natural Sciences and Engineering Research Council of Canada and the Earth and Life Sciences Foundation (ALW) of the Netherlands Organization for Scientific Research for ongoing research support. Thanks are also due to the following collaborators and students, including (in alphabetical order): Peter Bakker, Genrich Burd, Geneviève Défago, Varvara Grichko, Gina Holguin, Marga Knoester, Jiping Li, Wenbo Ma, Shimon Mayak, Jesus Mercado-Blanco, Barbara Moffatt, Seema Nayani, Daniel Ovakim, Peter Pauls, Donna Penrose, Corné Pieterse, Mary Robison, Saleh Shah, Steven Spoel, Jurriaan Ton, Ientse Van Der Sluis, Hans Van Pelt, Saskia Van Wees and Chunxia Wang.

References

Abeles FB (1973) Ethylene in plant biology. Academic Press, New York

Abeles FB, Morgan PW, Saltveit ME Jr (1992) Ethylene in plant biology, 2nd edn. Academic Press, San Diego

Alström S (1995) Evidence of disease resistance induced by rhizosphere pseudomonad against *Pseudomonas syringae* pv. *phaseolicola*. J Gen Appl Microbiol 41:315–325

Bakker PAHM, Ran LX, Pieterse CMJ, Van Loon LC (2003) Understanding the involvement of rhizobacteria-mediated induction of systemic resistance in biocontrol of plant diseases. Can J Plant Pathol 25:5–9

Bashan Y (1994) Symptom expression and ethylene production in leaf blight of cotton caused by *Alternaria macrospora* and *Alternaria alternata* alone and combined. Can J Bot 72:1574–1579

Belimov AA, Safronova VI, Sergeyeva TA, Egorova TN, Matveyeva VA, Tsyganov VE, Borisov AY, Tikhonovich IA, Kluge C, Preisfeld A, Dietz KJ, Stepanok VV (2001) Characterization of plant growth promoting rhizobacteria isolated from polluted soils and containing 1-aminocyclopropane-1-carboxylate deaminase. Can J Microbiol 47:642–652

Biles CL, Abeles FB, Wilson CL (1990) The role of ethylene in anthracnose of cucumber, *Cucumis sativus*, caused by *Colletotrichum lagenarium*. Phytopathology 80:732–736

Bostock RM (1999) Signal conflicts and synergies in induced resistance to multiple attackers. Physiol Mol Plant Pathol 55:99–109

Bradford KJ, Yang SF (1980) Xylem transport of 1-aminocyclopropane-1-carboxylic acid, an ethylene precursor, in waterlogged tomato plants. Plant Physiol 65:322–326

Brown ME (1974) Seed and root bacterization. Annu Rev Phytopathol 12:181–197

Burd GI, Dixon DG, Glick BR (1998) A plant growth-promoting bacterium that decreases nickel toxicity in seedlings. Appl Environ Microbiol 64:3663–3668

Campbell BG, Thomson JA (1996) 1-Aminocyclopropane-1-carboxylate deaminase genes from *Pseudomonas* strains. FEMS Microbiol Lett 138:207–210

Cohen R, Riov J, Lisker N, Katan J (1986) Involvement of ethylene in herbicide-induced resistance to *Fusarium oxysporum* f. sp. *melonis*. Phytopathology 76:1281–1285

Cronshaw DK, Pegg GF (1976) Ethylene as a toxin synergist in Verticillium wilt of tomato. Physiol Plant Pathol 9:33–38

Davison J (1988) Plant beneficial bacteria. Bio/Technology 6:282–286

De Meyer G, Höfte M (1997) Salicylic acid produced by the rhizobacterium *Pseudomonas aeruginosa* 7NSK2 induces resistance to leaf infection by *Botrytis cinerea* on bean. Phytopathology 87:588–593

De Meyer G, Audenaert K, Höfte M (1999) *Pseudomonas aeruginosa* 7NSK2-induced systemic resistance in tobacco depends on in planta salicylic acid accumulation but is not associated with PR1a expression. Eur J Plant Pathol 105:513–517

Duijff BJ, Pouhair D, Olivain C, Alabouvette C, Lemanceau P (1998) Implication of systemic induced resistance in the suppression of fusarium wilt of tomato by *Pseudomonas fluorescens* WCS417r and by non-pathogenic *Fusarium oxysporum* Fo47. Eur J Plant Pathol 104:903–910

Durner J, Shah J, Klessig DF (1997) Salicylic acid and disease resistance in plants. Trends Plant Sci 2:266–274

Elad Y (1988) Involvement of ethylene in the disease caused by *Botrytis cinerea* on rose and carnation flowers and the possibility of control. Ann Appl Biol 113:589–598

Elad Y (1990) Production of ethylene in tissues of tomato, pepper, French-bean and cucumber in response to infection by *Botrytis cinerea*. Physiol Mol Plant Pathol 36:277–287

Else MA, Jackson MB (1998) Transport of 1-aminocyclopropane-1-carboxylic acid (ACC) in the transpiration stream of tomato (*Lycopersicon esculentum*) in relation to foliar ethylene production and petiole epinasty. Aust J Plant Physiol 25:453–458

Farmer EE, Ryan CA (1992) Octadecanoid precursors of jasmonic acid activate the synthesis of wound-inducible proteinase inhibitors. Plant Cell 4:129–134

Felton GW, Korth KL (2000) Trade-offs between pathogen and herbivore resistance. Curr Opin Plant Biol 3:309–314

Felton GW, Korth KL, Bi JL, Wesley SV, Huhman DV, Mathews MC, Murphy JB, Lamb C, Dixon RA (1999) Inverse relationship between systemic resistance of plants to microorganisms and to insect herbivory. Curr Biol 9:317–320

Frankenberger WTJ, Arshad M (1995) Phytohormones in soil. Marcel Dekker, New York

Fukuda H, Ogawa T, Tanase S (1993) Ethylene production by microorganisms. Adv Microbial Physiol 35:275–306

Gaudin V, Vrain T, Jouanin L (1994) Bacterial genes modifying hormonal balances in plants. Plant Physiol Biochem 32:11–29

Giovanelli J, Mudd SH, Datko AH (1980) Sulfur amino acids in plants. In: Miflin BJ (ed) Amino acids and derivatives. The biochemistry of plants: a comprehensive treatise, vol 5. Academic Press, New York, pp 453–505

Glandorf DCM, Peters LGL, Van der Sluis I, Bakker PAHM, Schippers B (1993) Crop specificity of rhizosphere pseudomonads and the involvement of root agglutinins. Soil Biol Biochem 25:981–989

Glick BR (1995) The enhancement of plant growth by free-living bacteria. Can J Microbiol 41:109–117

Glick BR, Bashan Y (1997) Genetic manipulation of plant growth-promoting bacteria to enhance biocontrol of fungal phytopathogens. Biotechnol Adv 15:353–378

Glick BR, Jacobson CB, Schwarze MK, Pasternak JJ (1994) 1-Aminocyclopropane-1-carboxylic acid deaminase mutants of the plant growth promoting rhizobacterium *Pseudomonas putida* GR12-2 do not stimulate canola root elongation. Can J Microbiol 40:911–915

Glick BR, Karaturovíc DM, Newell PC (1995) A novel procedure for rapid isolation of plant growth promoting pseudomonads. Can J Microbiol 41:533–536

Glick, BR, Penrose DM, Li J (1998) A model for the lowering of plant ethylene concentrations by plant growth-promoting bacteria. J Theor Biol 190:63–68

Glick BR, Patten CL, Holguin G, Penrose, DM (1999). Biochemical and genetic mechanisms used by plant growth-promoting bacteria. Imperial College Press, London

Gómez-Gómez L, Boller T (2000) FLS2: an LRR receptor-like kinase involved in the perception of the bacterial elicitor flagellin in *Arabidopsis*. Mol Cell 5:1003–1011

Gómez-Gómez L, Felix G, Boller T (1999) A single locus determines sensitivity to bacterial flagellin in *Arabidopsis thaliana*. Plant J 18:277–284

Grichko VP, Glick BR (2000) Identification of DNA sequences that regulate the expression of the *Enterobacter cloacae* UW4 1-aminocyclopropane-1-carboxylate deaminase gene. Can J Microbiol 46:1159–1165

Grichko VP, Glick BR (2001) Amelioration of flooding stress by ACC deaminase-containing plant growth-promoting bacteria. Plant Physiol Biochem 39:11–17

Guinel FC, Sloetjes LL (2000) Ethylene is involved in the nodulation phenotype of *Pisum sativum* R50 (*sym16*), a pleiotropic mutant that nodulates poorly and has pale green leaves. J Exp Bot 51:885–894

Hall JA, Peirson D, Ghosh S, Glick BR (1996) Root elongation in various agronomic crops by the plant growth promoting rhizobacterium *Pseudomonas putida* GR12-2. Isr J Plant Sci 44:37–42

Hammerschmidt R, Métraux JP, Van Loon LC (2001) Inducing resistance: a summary of papers presented at the first international symposium on induced resistance to plant diseases, Corfu, May 2000. Eur J Plant Pathol 107:1–6

Handelsman J, Stabb EV (1996) Biocontrol of soilborne plant pathogens. Plant Cell 8:1855–1869

Höfte M, Bigirimana J, De Meyer G, Audenaert K (2000) Induced systemic resistance in tomato, tobacco and bean by Pseudomonas aeruginosa 7NSK2: bacterial determinants, signal transduction pathways and role in host resistance. Oral Sessions of the 5th International Workshop on PGPR, Cordoba, Argentina, pp 108–113

Hoffland E, Pieterse CMJ, Bik L, Van Pelt JA (1995) Induced systemic resistance in radish is not associated with accumulation of pathogenesis-related proteins. Physiol Mol Plant Pathol 46:309–320

Holguin G, Glick BR (2001) Expression of the ACC deaminase gene from *Enterobacter cloacae* UW4 in *Azospirillum brasilense*. Microbial Ecol 41:281–288

Honma M (1985) Chemically reactive sulfhydryl groups of 1-aminocyclopropane-1-carboxylate deaminase. Agric Biol Chem 49:567–571

Honma M (1993) Stereospecific reaction of 1-aminocyclopropane-1-carboxylate deaminase. In: Pech JC, Latché A, Balagué C (eds) Cellular and molecular aspects of the plant hormone ethylene. Kluwer, Dordrecht, pp 111–116

Honma M, Shimomura T (1978) Metabolism of 1-aminocyclopropane-1-carboxylic acid. Agric Biol Chem 42:1825–1831

Hyodo H (1991) Stress/wound ethylene. In: Mattoo AK, Suttle JC (eds) The plant hormone ethylene. CRC Press, Boca Raton, pp 65–80

Jacobson CB, Pasternak JJ, Glick BR (1994) Partial purification and characterization of 1-aminocyclopropane-1-carboxylate deaminase from the plant growth promoting rhizobacterium *Pseudomonas putida* GR12–2. Can J Microbiol 40:1019–1025

Jia YJ, Kakuta Y, Sugawara M, Igarashi T, Oki N, Kisaki M, Shoji T, Kanetuna Y, Horita T, Matsui H, Honma M (1999) Synthesis and degradation of 1-aminocyclopropane-1-carboxylic acid by *Penicillium citrinum*. Biosci Biotechnol Biochem 63:542–549

John P (1991) How plant molecular biologists revealed a surprising relationship between two enzymes, which took an enzyme out of a membrane where it was not located, and put it into the soluble phase where it could be studied. Plant Mol Biol Reporter 9:192–194

Kende H (1989) Enzymes of ethylene biosynthesis. Plant Physiol 91:1–4

Kende H (1993) Ethylene biosynthesis. Annu Rev Plant Physiol Plant Mol Biol 44:283–307

Klee HJ, Kishore GM (1992) Control of fruit ripening and senescence in plants. US patent number 5,702,933

Klee HJ, Hayford MB, Kretzmer KA, Barry GF, Kishore GM (1991) Control of ethylene synthesis by expression of a bacterial enzyme in transgenic tomato plants. Plant Cell 3:1187–1193

Kloepper JW, Lifshitz R, Zablotowicz RM (1989) Free-living bacterial inocula for enhancing crop productivity. Trends Biotechnol 7:39–43

Kloepper JW, Tuzun S, Kuc JA (1992) Proposed definitions related to induced disease resistance. Biocontrol Sci Technol 2:349–351

Knoester M, Pieterse CMJ, Bol JF, Van Loon LC (1999) Systemic resistance in Arabidopsis induced by rhizobacteria requires ethylene-dependent signaling at the site of application. Mol Plant–Microbe Interact 12:720–727

Koike N, Hyakumachi M, Kageyama K, Tsuyumu S, Doke N (2001) Induction of systemic resistance in cucumber against several diseases by plant growth-promoting fungi: lignification and superoxide generation. Eur J Plant Pathol 107:523–533

Lambert B, Joos H (1989) Fundamental aspects of rhizobacterial plant growth promotion research. Trends Biotechnol 7:215–219

Leeman M, Van Pelt JA, Den Ouden FM, Heinsbroek M, Bakker PAHM, Schippers B (1995a) Induction of systemic resistance by *Pseudomonas fluorescens* in radish cultivars differing in susceptibility to fusarium wilt, using a novel bioassay. Eur J Plant Pathol 101:655–664

Leeman M, Van Pelt JA, Den Ouden FM, Heinsbroek M, Bakker PAHM, Schippers B (1995b) Induction of systemic resistance against fusarium wilt of radish by lipopolysaccharides of *Pseudomonas fluorescens*. Phytopathology 85:1021–1027

Leeman M, Den Ouden FM, Van Pelt JA, Dirkx FPM, Steijl H, Bakker PAHM, Schippers B (1996) Iron availability affects induction of systemic resistance against fusarium wilt of radish by *Pseudomonas fluorescens*. Phytopathology 86:149–155

Li J, Glick BR (2001) Transcriptional regulation of the *Enterobacter cloacae* UW4 1-aminocyclopropane-1-carboxylate (ACC) deaminase gene (*acdS*). Can J Microbiol 47:259–267

Li J, Ovakim D, Charles TC, Glick BR (2000) An ACC deaminase minus mutant of *Enterobacter cloacae* UW4 no longer promotes root elongation. Curr Microbiol 41:101–105

Lund ST, Stall RE, Klee HJ (1998) Ethylene regulates the susceptible response to pathogen infection in tomato. Plant Cell 10:371–382

Lynch JM, Whipps JM (1991) Substrate flow in the rhizosphere. In: Keister DL, Cregan PB (eds) The rhizosphere and plant growth. Kluwer, Dordrecht, pp 15–24

Ma W, Sebestianova SB, Sebestian J, Burd GL, Guinel FC, Glick BR (2003) Prevalence of 1-aminocyclopropane-1-carboxylate deaminase in *Rhizobium* spp. Antonie von Leeuwenhoek 83:285–291

Maleck K, Levine A, Eulgem T, Morgan A, Schmid J, Lawton KA, Dangl JL, Dietrich RA (2000) The transcriptome of *Arabidopsis thaliana* during systemic acquired resistance. Nat Genet 26:403–410

Mattoo AK, Suttle JC (eds) (1991) The plant hormone ethylene. CRC Press, Boca Raton

Maurhofer M, Hase C, Meuwly P, Métraux JP, Défago G (1994) Induction of systemic resistance of tobacco to tobacco necrosis virus by the root-colonizing *Pseudomonas fluorescens* strain CHA0: influence of the *gacA* gene and of pyoverdine production. Phytopathology 84:139–146

Meyer JM, Azelvandre P, Georges C (1992) Iron metabolism in *Pseudomonas*: salicylic acid, a siderophore of *Pseudomonas fluorescens* CHA0. BioFactors 4:23–27

Minami R, Uchiyama K, Murakami T, Kawai J, Mikami K, Yamada T, Yokoi D, Ito H, Matsui H, Honma M (1998) Properties, sequence, and synthesis in *Escherichia coli* of 1-aminocyclopropane-1-carboxylate deaminase from *Hansenula saturnus*. J Biochem 123:1112–1118

Mol JNM, Holton TA, Koes RE (1995) Floriculture: genetic engineering of commercial traits. Trends Biotechnol 13:350–355

Morgan PW, Drew CD (1997) Ethylene and plant responses to stress. Physiol Plant 100:620–630

Murphy AM, Chivasa S, Singh DP, Carr JP (1999) Salicylic acid-induced resistance to viruses and other pathogens: a parting of the ways? Trends Plant Sci 4:155–160

Naseby DC, Lynch JM (1998) Establishment and impact of *Pseudomonas fluorescens* genetically modified for lactose utilization and kanamycin resistance in the rhizosphere of pea. J Appl Microbiol 84:169–175

Nayani S, Mayak S, Glick BR (1998) The effect of plant growth promoting rhizobacteria on the senescence of flower petals. Ind J Exp Biol 36:836–839

O'Connell KP, Goodman RM, Handelsman J (1996) Engineering the rhizosphere: expressing a bias. Trends Biotechnol 14:83–88

Patten CL, Glick BR (1996) Bacterial biosynthesis of indole-3-acetic acid. Can J Microbiol 42:207–220

Paul ND, Hatcher PE, Taylor JE (2000) Coping with multiple enemies: an integration of molecular and ecological perspectives. Trends Plant Sci 5:220–225

Penmetsa RV, Cook DR (1997) A legume ethylene-insensitive mutant hyperinfected by its rhizobial symbiont. Science 275:527–530

Penrose DM, Glick BR (2001) Levels of ACC and related compounds in exudate and extracts of canola seeds treated with ACC deaminase-containing plant growth-promoting bacteria. Can J Microbiol 47:368–372

Pieterse CMJ, Van Loon LC (1999) Salicylic acid-independent plant defence pathways. Trends Plant Sci 4:52–58

Pieterse CMJ, Van Wees SCM, Hoffland E, Van Pelt JA, Van Loon LC (1996) Systemic resistance in *Arabidopsis* induced by biocontrol bacteria is independent of salicylic acid accumulation and pathogenesis-related gene expression. Plant Cell 8:1225–1237

Pieterse CMJ, Van Wees SCM, Van Pelt JA, Knoester M, Laan R, Gerrits H, Weisbeek PJ, Van Loon LC (1998) A novel signaling pathway controlling induced systemic resistance in *Arabidopsis*. Plant Cell 10:1571–1580

Pieterse CMJ, Van Pelt, JA, Ton J, Parchmann S, Mueller MJ, Buchala AJ, Métraux JP, Van Loon LC (2000) Rhizobacteria-mediated induced systemic resistance (ISR) in *Arabidopsis* requires sensitivity to jasmonate and ethylene but is not accompanied by an increase in their production. Physiol Mol Plant Pathol 57:123–134

Raaijmakers JM, Leeman M, Van Oorschot MPM, Van der Sluis I, Schippers B, Bakker PAHM (1995) Dose-response relationships in biological control of fusarium wilt of radish by *Pseudomonas* spp. Phytopathology 85:1075–1081

Reymond P, Weber H, Damond M, Farmer EE (2000) Differential gene expression in response to mechanical wounding and insect feeding in *Arabidopsis*. Plant Cell 12:707–719

Robison MM (2001) Dual role for ethylene in susceptibility of tomato to *Verticillium* wilt. J Phytopathol 149:385–388

Ryals JA, Neuenschwander UH, Willits MG, Molina A, Steiner HY, Hunt MD (1996) Systemic acquired resistance. Plant Cell 8:1809–1819

Ryan CA (1992) The search for the proteinase-inhibitor inducing factor, PIIF. Plant Mol Biol 19:123–133

Schenk PM, Kazan K, Wilson I, Anderson JP, Richmond T, Somerville SC, Manners JM (2000) Coordinated plant defense responses in *Arabidopsis* revealed by microarray analysis. Proc Natl Acad Sci USA 97:11655–11660

Schippers B, Bakker AW, Bakker PAHM (1987) Interactions of deleterious and beneficial rhizosphere micro-organisms and the effect of cropping practices. Annu Rev Phytopathol 25:339–358

Shah S, Li J, Moffatt BM, Glick BR (1997) ACC deaminase genes from plant growth promoting bacteria. In: Ogoshi A, Kobayashi K, Homma Y, Kodama F, Kondo N, Akino S (eds) Plant growth-promoting rhizobacteria: present status and future prospects. OECD, Paris, pp 320–324

Shah S, Li J, Moffatt BA, Glick BR (1998) Isolation and characterization of ACC deaminase genes from two different plant growth promoting rhizobacteria. Can J Microbiol 44:833–843

Sheehy RE, Honma M, Yamada M, Sasaki T, Martineau B, Hiatt WR (1991) Isolation, sequence, and expression in *Escherichia coli* of the *Pseudomonas* sp. strain ACP gene encoding 1-aminocyclopropane-1-carboxylate deaminase. J Bacteriol 173:5260–5265

Sisler EC, Serek M (1997) Inhibitors of ethylene responses in plants at the receptor level: recent developments. Physiol Plant 100:577–582

Spoel SH, Koornneef A, Claessens SMC, Korzelius JP, van Pelt JA, Mueller MJ, Buchala AJ, Métraux JP, Brown R, Kazan K, Van Loon LC, Dong X, Pieterse CMJ (2003) NPR1 modulates cross-talk between salicylate- and jasmonate-dependent defense pathways through a novel function in the cytosol. Plant Cell 15:760–770

Thomma BPHJ, Eggermont K, Penninckx IAMA, Mauch-Mani B, Vogelsang R, Cammue BPA, Broekaert WF (1998) Separate jasmonate-dependent and salicylate-dependent defense response pathways in *Arabidopsis* are essential for resistance to distinct microbial pathogens. Proc Natl Acad Sci USA 95:15107–15111

Thomma BPHJ, Penninckx IAMA, Broekaert WF, Cammue BPA (2001a) The complexity of disease signaling in *Arabidopsis*. Curr Opin Immunol 13:63–68

Thomma BPHJ, Tierens KFM, Penninckx IAMA, Mauch-Mani B, Broekaert WF, Cammue BPA (2001b) Different micro-organisms differentially induce Arabidopsis disease response pathways. Plant Physiol Biochem 39:673–680

Thulke OU, Conrath U (1998) Salicylic acid has a dual role in the activation of defence-related genes in parsley. Plant J 14:35–43

Ton J, Pieterse CMJ, Van Loon LC (1999) Identification of a locus in Arabidopsis control-ling both the expression of rhizobacteria-mediated induced systemic resistance (ISR) and basal resistance against *Pseudomonas syringae* pv. *tomato*. Mol Plant–Microbe Interact 12:911–918

Ton J, Davison S, Van Wees SCM, Van Loon LC, Pieterse CMJ (2001) The Arabidopsis *ISR1* locus controlling rhizobacteria-mediated induced systemic resistance is involved in ethylene signaling. Plant Physiol 125:652–661

Ton J, Van Pelt JA, Van Loon, LC, Pieterse CMJ (2002a) Differential effectiveness of sali-cylate-dependent and jasmonate/ethylene-dependent induced resistance in *Ara-bidopsis*. Mol Plant–Microbe Interact 15:27–34

Ton J, De Vos M, Robben C, Buchala AJ, Métraux JP, Van Loon LC, Pieterse CMJ (2002b) Characterisation of *Arabidopsis* enhanced disease susceptibility mutants that are affected in systemically induced resistance. Plant J 29:11–21

Ton J, Van Pelt JA, Van Loon LC, Pieterse CMJ (2002c) The Arabidopsis *ISR1* locus is required for rhizobacteria-mediated induced systemic resistance against different pathogens. Plant Biol 4:224–227

Troxler J, Berling CH, Moënne-Loccoz Y, Keel C, Défago G (1997) Interactions between the biocontrol agent *Pseudomonas fluorescens* CHA0 and *Thielaviopsis basicola* in tobacco roots observed by immunofluorescence microscopy. Plant Pathol 46:62–71

Tuzun S, Kloepper J (1995) Practical application and implementation of induced resis-tance. In: Hammerschmidt R, Kuć J (eds) Induced resistance to disease in plants. Kluwer, Dordrecht, pp 152–168

Van Loon LC (1984) Regulation of pathogenesis and symptom expression in diseased plants by ethylene. In: Fuchs Y, Chalutz E (eds) Ethylene: biochemical, physiological and applied aspects. Martinus Nijhoff/Dr W. Junk, The Hague, pp 171–180

Van Loon LC (1997) Induced resistance in plants and the role of pathogenesis-related proteins. Eur J Plant Pathol 103:753–765

Van Loon LC (2000) Systemic induced resistance. In: Slusarenko AJ, Fraser RSS, Van Loon LC (eds) Mechanisms of resistance to plant diseases. Kluwer, Dordrecht, pp 521–574

Van Loon LC, Pieterse CMJ (2002) Biocontrol agents in signaling resistance. In: Gnana-manickam SS (ed) Biological control of crop diseases. Marcel Dekker, New York, pp 355–386

Van Loon LC, Bakker PAHM, Pieterse CMJ (1998) Systemic resistance induced by rhizo-sphere bacteria. Annu Rev Phytopathol 36:453–483

Van Peer R, Schippers B (1992) Lipopolysaccharides of plant-growth promoting *Pseudomonas* sp. strain WCS417r induce resistance in carnation to fusarium wilt. Neth J Plant Pathol 98:129–139

Van Peer R, Niemann GJ, Schippers B (1991) Induced resistance and phytoalexin accu-mulation in biological control of fusarium wilt of carnation by *Pseudomonas* sp. strain WCS417r. Phytopathology 81:728–734

Van Wees SCM, Pieterse CMJ, Trijssenaar A, Van 't Westende Y, Hartog F, Van Loon LC (1997) Differential induction of systemic resistance in *Arabidopsis* by biocontrol bac-teria. Mol Plant–Microbe Interact 10:716–724

Van Wees SCM, Luijendijk M, Smoorenburg I, Van Loon LC, Pieterse CMJ (1999) Rhi-zobacteria-mediated induced systemic resistance (ISR) in *Arabidopsis* is not associated with a direct effect on expression of known defense-related genes but stimulates the expression of the jasmonate-inducible gene *Atvsp* upon challenge. Plant Mol Biol 41:537–549

Van Wees SCM, De Swart EAM, Van Pelt JA, Van Loon LC, Pieterse CMJ (2000) Enhancement of induced disease resistance by simultaneous activation of salicylate- and jasmonate-dependent defense pathways in *Arabidopsis thaliana*. Proc Natl Acad Sci USA 97:8711–8716

Walling LL (2000) The myriad plant responses to herbivores. J Plant Growth Regul 19:195–216

Walsh UF, Morrissey JP, O'Gara F (2001) *Pseudomonas* for biocontrol of phytopathogens: from functional genomics to commercial exploitation. Curr Opin Biotechnol 12:289–295

Wang C, Knill E, Glick BR, Défago G (2000) Influence of 1-aminocyclopropane-1-carboxylic acid (ACC) deaminase genes transferred into *Pseudomonas fluorescens* strain CHA0 and its *gacA* derivative CHA96 on their growth-promoting and disease-suppressive capacities. Can J Microbiol 46:898–907

Wei G, Kloepper JW, Tuzun S (1991) Induction of systemic resistance of cucumber to *Colletotrichum orbiculare* by select strains of plant growth-promoting rhizobacteria. Phytopathology 81:1508–1512

Whipps JM (1990) Carbon utilization. In: Lynch JM (ed) The rhizosphere. Wiley Interscience, Chichester, UK, pp 59–97

Woltering EJ, Van Doorn WG (1988) Role of ethylene in senescence of petals – morphological and taxonomical relationships. J Exp Bot 39:1605–1616

Yang SF, Hoffman NE (1984) Ethylene biosynthesis and its regulation in higher plants. Annu Rev Plant Physiol 35:155–189

Zehnder GW, Murphy JF, Sikora EJ, Kloepper JW (2001) Application of rhizobacteria for induced resistance. Eur J Plant Pathol 107:39–50

8 Scaling Up from Molecular to Ecological Processes

H. SANDERMANN and R. MATYSSEK

8.1 Overview

Previous chapters in this book have described a number of case studies whose contributions to the concept outlined in Chapter 1 are summarized in Table 8.1. Environmental stress is shown to be countered by numerous pre-formed or induced genetic and biochemical responses. These may be used operationally as biomarkers to characterize partial aspects of plant fitness. A note of caution is obligatory – so far, this use of molecular biomarkers has only been documented in a limited number of laboratory studies. Development of biomarkers for field use is clearly a task for the future, and current problems of plant and site variability have to be overcome. In addition, functional genomic and biochemical studies have led to a multidimensional view of plant signalling and stress-response networks that has been depicted in the form of a Venn diagram in Chapter 1 (Fig. 1.5). On this basis, it is predicted that the elucidation of signalling and stress response networks in a systematic way will be a major focus of future research.

Examples for such demanding approaches can be cited in addition to the case studies of Table 8.1. For example, studies of herbivory have been successful in defining regulatory networks showing similarities and differences with regard to mechanical stress (wounding) and pathogen attack (Kessler and Baldwin 2002). A second example has uncovered surprising similarities between innate immunity and certain virulence mechanisms in plants, animals and humans (Cao et al. 2001; Gómez-Gómez and Boller 2002; Janeway and Medzhitov 2002). The ecological risk assessment of transgenic plants represents an additional emerging field of molecular research. This politically urgent practical application of molecular ecotoxicology is briefly outlined in Text Box 1.

The outlined research developments will require concepts for scaling up from molecular to ecological processes. Scaling up may proceed in a case-specific manner, but general concepts to integrate molecular methods into ecological research are desirable. Two such approaches will be described in the following text, namely ecophysiology and molecular epidemiology.

Ecological Studies, Vol. 170
H. Sandermann (Ed.)
Molecular Ecotoxicology of Plants
© Springer-Verlag Berlin Heidelberg 2004

Table 8.1. Contributions of individual book chapters to the concept of molecular ecotoxicology outlined in Chapter 1

Chapter, keywords	Input	Response networks	Output
Chapter 2, glutathione S-transferases; detoxification; natural substrates; xenobiotics; safeners; stress defense	Induction by biotic and abiotic stress, by chemicals	Gene induction; interaction with AOS; promoter structure	Detoxification; vacuolar deposition; herbicide selectivity
Chapter 3, activated oxygen species/endogenous and environmental induction; multiple stresses; biotic/abiotic interaction	AOS production under endogenous and environmental influences	Signalling networks, biotic/abiotic interaction	Adaptation or damage; induced resistance; adaptation to desert field site
Chapter 4, programmed cell death, hypersensitive response, biosynthesis of jasmonic acid, of ethylene	Pathogen attack; abiotic stress; ozone as pollutant	Cell death programs; signalling chains; jasmonic acid, ethylene as second messengers	Induced resistance, necrotic death
Chapter 5, programmed cell death, induced resistance, systemic acquired resistance; hypersensitive response; biosynthesis of salicylic acid, of NO	Pathogen attack, abiotic stress, NO as pollutant	Salicylic acid, NO as second messengers; signalling networks; priming	Induced resistance
Chapter 6, gene promoters, cis elements, transcription factors	Temperature, water stress; UV-B; ozone; elicitors; xenobiotics; second messengers	Gene activation; protein/DNA-interactions	Repair and defense proteins; protective metabolites
Chapter 7, plant fitness/rhizobacteria/ abiotic and biotic stress in the rhizosphere	Abiotic stress, pathogen attack in rhizosphere	Ethylene downregulation; bacterial signals	Increased fitness; induced resistance, growth promotion

Text Box 1
Ecological Questions Posed by Transgenic Plants

Transgenic plants are presently cultivated commercially on some ~45 million ha, much of it in the USA. In 1989, The Ecological Society of America published a comprehensive risk assessment (Tiedje et al. 1989), but in 1995–1996, the US Government went ahead with deregulation and large-scale commercialization without establishing testing and post-release monitoring protocols. This has resulted in distinct criticism with adjectives such as "flawed" (National Research Council 2000) and "superficial" (National Research Council 2002). At present, both the Cartagena Protocol on Biosafety (2001) and the EU guideline 2001/18/EC call for risk evaluation of transgenic plants on a scientific basis and for post-release monitoring. A huge area of research for molecular ecotoxicology is thereby opened up, in particular because most transgenic crop lines commercialized so far carry resistance traits that could change plant susceptibility to environmental stress. It is not known whether the huge soil input of plant-associated antibiotic resistance genes (several kilograms plant DNA per hectare are typically released per growing season) will change the already existing microbial soil pool of such genes. Several studies have indicated that the lifetime of plant genes in soils is in the order of weeks and months. With regard to herbicide resistance, the prevailing broad-spectrum herbicides glyphosate or Basta are also active as antibiotics (Sandermann 1997), but effects on soil microbial biodiversity, weed resistance and cross-tolerances (cf. Chap. 1) have hardly been studied. With regard to insect resistance, strategies to avoid the evolution of resistant insects under the selection pressure of transgenic Bt-crops are still disputed. To what extent incorporated foreign genes including viral DNA will undergo silencing or recombination events is also not well understood. These processes could result in virus inactivation or in the generation of new pathogens. In summary, systematic research to link the genotype of transgenic plants to their ecological phenotypes as well as processes of gene flow, selection pressure and population dynamics is urgently needed (see Sandermann 2003).

8.2 Ecophysiology

The role of plants as primary producers in ecosystems is briefly summarized in Fig. 8.1. The numerous participating processes and interactions need to be analyzed at the molecular level, i.e., of gene transcription, translation and post-translational regulation (for example, metabolite concentrations and fluxes). These basic processes need to be considered across a hierarchy of organizational levels, as illustrated in Fig. 8.2. Biological organization comprises the molecular processes, the structural and physiological coordination within and between cells and tissues, the performance of plant organs and the functioning of the entire plant. The plant itself may be viewed as part of populations and ecosystems (cf. Matyssek et al. 1995), and the latter represent constituents within landscapes and the biogeochemical cycles of the entire globe (Schlesinger 1997). The functional properties of each level have a component in space and time and may be conceived as integrated into the functional behavior of the adjacent higher level of the hierarchy. Processes interact

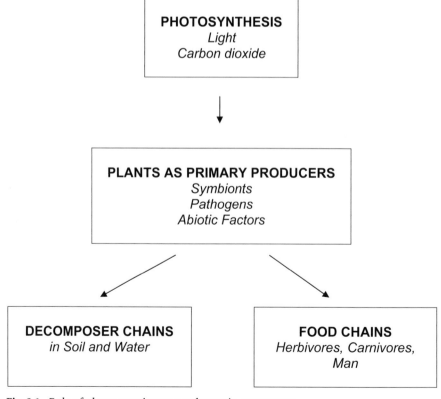

Fig. 8.1. Role of plants as primary producers in ecosystems

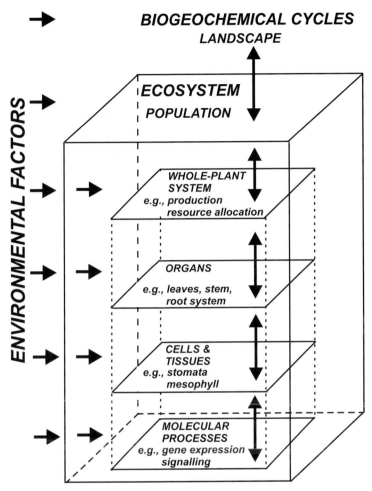

Fig. 8.2. Hierarchical levels of biological organization in structure and function. The interactions between the levels represent the driving forces of the processes occurring within each level. Environmental factors may act on each of these levels, whereas impact on any level may indirectly influence processes on all other levels. Scaling procedures may reach across such levels of different spatio-temporal complexity (see text for details; adapted from Matyssek et al. 1995)

between the levels, and each level is under the influence of the abiotic and biotic environment. Given this network of interaction, scaling means the analysis of the impact of processes at the one level on the performance of processes within adjacent levels.

There are general principles and rules in scaling, which were introduced in excellent accounts in the 1990s (Ehleringer and Field 1993; van Gardingen et al. 1997a). First of all, one must not expect individual, integrative processes to exist that would bridge plant functioning between the molecular and the whole-plant levels, perhaps to be expressed as one discrete function in mathematical modeling, or even to extend to the higher hierarchical levels. Rather, the higher the level, the less the interactions turn to being stoichiometric while becoming more variable; non-linearity of interactions increases, structural dimensions become larger, as does their interface with the impacting environment, and assessments need progressively to be conducted under a long-term rather than short-term perspective (Levin 1993). Hence, proceeding from the molecular to the higher levels, i.e., pursuing the "bottom-up" approach in functional integration (Jarvis 1993), complexity increases, and statistical rather than experimental analysis gains in importance. This does not mean that findings at high hierarchical levels are less mechanistic in their functional properties than molecular processes, which are sometimes claimed to be the only mechanistic ones; rather, the spatio-temporal resolution of mechanisms obeys distinct rules along the hierarchical levels shown in Fig. 8.2.

Given the above rules, scaling of processes, i.e., the analysis of process coupling across hierarchy levels, must be restricted to the interaction between three adjacent levels (Baldocchi 1993): the reductionist, operational and macro-levels. Mechanisms of the operational level need to be explained in functional terms on the basis of those at the underlying reductionist level, and the factors driving the mechanisms at the operational (and reductionist) level are determined at the macro level. In Fig. 8.2, this is exemplified through the molecular (reductionist), cell/tissue (operational) and organ (macro) levels – or cell/tissue, organ and whole-plant levels – or organ, whole-plant and ecosystem levels. Bottom-up scaling of ecophysiological processes typically starts at the organ level (Kruijt et al. 1997), whereas the coupling of molecular mechanisms with processes of cell/tissue biochemistry and physiology (or even to those at higher levels) continues to be a challenge. Many studies have unraveled a broad spectrum of molecular and biochemical plant responses under laboratory conditions (see previous chapters). However, such highly resolved mechanisms can hardly be integrated to be directly applied to plant performance under "real" site conditions (Bazzaz 1997) because of the spatio-temporal variability of the multi-factorial influences within ecosystems. Therefore, incorporation of such basic responses into quantitative bottom-up scaling concepts that may reach the ecosystem level is not feasible because of the encountered complexity (Jarvis 1993; Kruijt et al. 1997). Nevertheless, knowledge of basic mechanisms is indispensable for the explanation of ecophysiological responses at the organ and whole-plant levels. Thus, ecophysiological processes may be regarded as "scaling units" that integrate plant performance at the lower hierarchical levels in conceptual terms while providing an interface and platform for up-scaling to the higher levels in the hierarchy (Reynolds et al. 1993; Kolb and Matyssek 2001).

A prerequisite of quantitative response scaling, e.g., through mathematical models, across hierarchical levels is the conceptual understanding of how processes are interrelated with each other. This has been exemplified by Kolb and Matyssek (2001) and extended by Matyssek and Sandermann (2003) for tree response to ozone (which can be regarded, to some extent, as a stress mimicking pathogen impact; Sandermann 1996) at the cell, organ and whole-plant levels. After uptake through the stomata and impact on the mesophyll apoplast, ozone (in particular, its oxidative derivatives) initiates oxidative burst reactions that promote metabolic signalling towards gene expression and subsequent, post-transcriptional stress responses. The one pathway initiates programmed cell death, formation of pathogenesis-related proteins and a spectrum of defense metabolites, which are measures that demand energy as reflected by raised respiration, increase in PEPC activity and repair processes at the leaf level as well as a "less negative" $\delta^{13}C$ signature in the stem cellulose. Secondary compounds may be formed that foster lignin synthesis and contribute to an enhanced leaf mass/area ratio. Another pathway initiates chloroplast decline and membrane injury that may result in the disturbance of cell water relations and ion fluxes and inhibit leaf extension growth while fostering leaf structural decline and loss. Related to chloroplast impairment and changed leaf differentiation are declines in Rubisco activitiy and photosynthesis as well as partial stomatal closure at increased stomata density. The responses at the leaf level, both through active metabolic regulation and injury, cause an altered resource allocation in the whole plant. The latter is reflected, e.g., in changed branching pattern, leaf turnover, reduced root/shoot biomass ratio and foliage area and is accompanied by a declining "water-use efficiency" of the carbon gain, declining reserve storage and fructification and eventually limited biomass production. Overall, an increased defense status appears to occur at the expense of other aspects of whole-plant fitness (e.g., competitiveness relative to neighboring plants or reproduction). Although the principles of process coupling are well understood across three levels of plant-internal organization in the case of ozone impact, it can be demonstrated that specific scenarios of abiotic and biotic impacts can modify the response patterns to ozone dramatically (Polle et al. 2000; Matyssek 2001a). Moreover, plant ontogeny itself is a driving factor in the plant's sensitivity to stress (Laurence et al. 1994; Kolb and Matyssek 2001). Therefore, further "dimensions" in process scaling need to be accounted for, aside from those addressed so far, i.e., concerning plant ontogeny and the transfer of findings from the laboratory to "real" site conditions (Matyssek 2001b).

The above example demonstrates that bottom-up scaling becomes questionable and even unfeasible towards high levels of process hierarchy (Jarvis 1993). The shortcoming of this approach is intrinsically related to the scaling rules pointed out above, but may be compensated for by combining "bottom-up" with "top-down" concepts (Kruijt et al. 1997). Such latter concepts attempt to unravel ecosystem performance by relating it to functional "bulk parameters" and breaking it up, in a descriptive way, into interacting compartments (e.g., canopy, root-soil system). The analysis can assign a "big leaf" behavior to ecosystems and be related to the resource allocation between the functional compartments (Jarvis 1993). On the other hand, resource allocation turned out – within bottom-up approaches – to be of crucial relevance for the understanding of whole-plant functioning (cf. above example on ozone stress). Thus, resource allocation can provide an adequate interface between bottom-up and top-down approaches in scaling. A basically similar research strategy of integrated analysis of plant ecophysiology at the levels listed in Table 8.2 has previously been recommended by Sandermann et al. (1997).

Hence, the regulation in resource allocation is one focal point within scaling schemes, while its functional understanding still demands particular efforts in plant ecophysiological research (Schulze 1994; Bazzaz and Grace 1997). Adjustment between the three major resource fluxes of carbon, water and nutrients in plants has been conceived as being mediated through some kind of balancing regulation between the carbon versus the other two fluxes. The target balance to be achieved was termed "set-point" (Mooney and Winner 1991; Matyssek et al. 1998), which refers to the regulatory homeostasis of the unstressed plant. Environmental influences tend to cause deflection from the set-point – a disturbance to be overcome by the plant by re-adjustment of its resource fluxes and, hence, allocation between physiological processes and organs. This regulation can be reflected, e.g., by the C/N or the root/shoot biomass ratio, provided such latter ratios do not merely mirror developmental trends during plant ontogeny (Müller et al. 2000). Stitt and Schulze (1994) have extended this concept in a meticulous way, relating the regulation between the carbon, water and nutrient fluxes in the whole plant to a molecular basis and demonstrating that Rubisco and nitrate reductase act as central players in this interaction. One must be aware, however, that resource allocation inherently shifts between primary and secondary metabolism in relation to varying resource availability, in particular of carbon or nutrients (Fig. 8.3).

Table 8.2. Experimental hierarchy proposed for ozone research. (Sandermann et al. 1997)

The plant as an individual:
Adult plant, seedling, cell culture: gene expression, stress metabolism. Acclimation or damage of vegetative and reproductive functions, including root/shoot allocations and primary/secondary metabolism

---

The plant as part of its microenvironment:
Effects of O_3 on mycorrhization, nutrient uptake and susceptibility towards pathogens, frost, heat or other microclimatic factors

---

The plant as part of an ecosystem:
Population genetics and shifts; gene flow. Food chains. Interactions with non-plant organisms including herbivores, saprophytes and decomposers. Element, water and carbon cycling

---

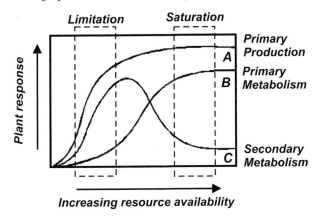

Fig. 8.3. Impact of resource availability on primary production as well as on primary and secondary metabolism (see text for details; adapted from Herms and Mattson 1992; Matyssek et al. 2002)

Increased primary production enables the plant to grow rapidly and reproduce amply, while out-competing neighboring plants in terms of above- and below-ground resource sequestration; in other words, competitiveness increases with respect to an efficient exploitation of the above- and below-ground space for resources (cf. Grams et al. 2002). In parallel, a decrease in secondary metabolism may occur. However, this leads to a restricted defense capacity against herbivores or pathogens. Whether this switching in allocation may be viewed as a trade-off – or even dilemma – of plants to grow or to defend (Herms and Mattson 1992) rather than an optimization principle for coping with different physiological needs is still an issue of ongoing research (Heil and Baldwin 2002; Matyssek et al. 2002). In any case, the trade-off hypothesized in Fig. 8.3 reflects the functional basis of individual plant fitness, i.e., the capacities in the sequestration (competitiveness) and retention (defense) of resources.

Molecular biomarkers to determine the status of plants in growth and defense are only beginning to become available. However, the validity of simplifying concepts like that of Fig. 8.3 still needs to be examined along the functional hierarchy shown in Table 8.2 and Fig. 8.2. Although some studies are consistent with this concept, others are not, because of a lack in mechanistic understanding. In particular, influences of regulatory factors such as symbionts (e.g., mycorrhizae), pathogens/herbivores and competing plants have been neglected, as have such abiotic factors as, e.g., water status, nutrition or ozone exposure, which altogether can bias or dominate the performance of plants and their stress response (see above account on tree response to ozone stress or debate of "global change" effects in plants: Saxe et al. 1998; Matyssek and Innes 1999). However, molecular mechanisms for these kinds of interaction are largely unknown, and biomarkers have not yet been defined.

This deficit in research is currently being addressed by an interdisciplinary research program conducted by 20 working groups in the Munich area entitled *Growth and Parasite Defense – Competition for Resources in Economic*

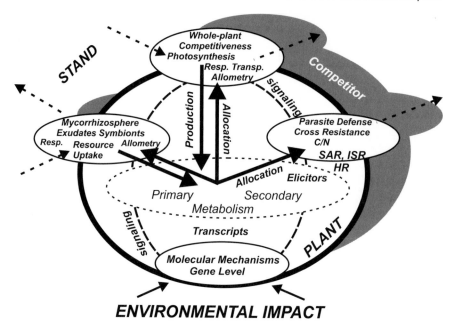

Fig. 8.4. Conceptual model underlying the interdisciplinary research program *SFB 607: Growth and Parasite Defense – Competition for Resources in Economic Plants from Agronomy and Forestry. Arrows* indicate major pathways in resource allocation between physiological demands within the plant and in exchange with the environment as being under the control of environmental impact and molecular processes (see text for details)

Plants from Agronomy and Forestry (Matyssek et al. 2002). The pursued concept as visualized in Fig. 8.4 focuses on the regulatory mechanisms of resource allocation between and within herbaceous and woody plants under the influence of plant competitors as well as symbiontic and parasitic organisms. Resource allocation is being analyzed as driven by host-parasite interactions, mycorrhizospheric relationships and competition with neighboring plants in stands, assessing the associated cost/benefit relationships of resource turnover and exchange with the environment and the "opportunity costs" of alternative resource uses within the metabolism. Hence, the primary competition for resources, before being extended to that among neighboring plants at the stand level, occurs plant-internally already between the different demands for the three major ranges of biotic interaction and within the resource pool (including storage) provided by the primary and secondary metabolism. This latter pool is – via transcriptional coupling – under the control of molecular processes, which in turn respond to the environmental impact. Signal transduction across the whole-plant system ensures that the plant is able to function both across its internal hierarchy levels of biological organization and across the three ranges of biotic interaction as an integral entity. The extent to which central mechanisms in the control of resource allocation behave similarly in woody and herbaceous plants, young and mature individuals, and under laboratory and site conditions needs to be clarified. Experimental concepts such as those depicted in Fig. 8.4 have the capacity to bridge the functional gap between molecular and ecophysiological processes. Proceeding with concepts of this kind will establish the knowledge needed at the molecular level with regard to the driving forces of natural successions or contrasting adaptations of different plant life forms and will provide a functional basis for geobotanical assessment of plant growth strategies (Grime et al. 1988; cf. Stitt and Schulze 1994). For example, the molecular analysis of growth strategies of competitors, ruderals and stress-tolerator plants will open a huge field for future research.

Do ecophysiological "biomarkers" already exist? Basically, yes, when thinking of plant parameters such as photosynthetic CO_2 uptake, chlorophyll fluorescence, respiration, stomatal conductance, root and shoot architecture or the adjustment between the fluxes in resource allocation and their induced signatures of stable isotopes in the plant biomass (Ehleringer et al. 1993). However, the response of such and other parameters to environmental stress can be rather variable, due to the increasing complexity of multi-factorial impacts at high hierarchical levels of biological organization – a fact being reflected by the scaling rules addressed above. A way for coping with this variability in plant response is the analysis of response patterns of plant parameters that are spread across several functional hierarchy levels within the plant (including the molecular one) and that respond in parallel to each other, rather than focusing on the behavior of single parameters without a context to the remainder of the whole-plant system. Hence, integrative response patterns to be tested for consistency rather than the responsiveness of isolated plant parameters may serve as biomarkers in ecophysiological terms. Such an approach is currently being conducted at Kranzberg Forest near Munich, where adult beech and spruce trees are analyzed for their competitiveness and defense capacities under the prevailing and experimentally enhanced, chronic "free-air" ozone exposure (Fig. 8.5; Nunn et al. 2002; Werner and Fabian 2002).

Today, there are ways available for closing the gap in the understanding of the functional coupling between molecular and ecophysiological processes. However, existing scaling concepts need to be extended, or even new ones need to be established, in order to achieve this goal. The presented account gives some perspectives in doing so. The challenge is to leave a strict "reductionist's perspective" in the research on basic processes, which biological science traditionally has been used to and has fostered towards increasing "specialization" (Grace et al. 1997) – and which may be a temptation also in the research field of molecular biology. Rather, the demanding task of interdisciplinary work between biological disciplines will gain in importance for unraveling general principles across the functional levels of biological systems as they interact with the "real world" of field sites, biogeochemical cycles

Fig. 8.5. (in color at end of book) Experimental site "Kranzberg Forest" (near Munich, Germany) for studying mechanisms of competitiveness and ozone sensitivity in adult beech and spruce trees. Vertical tubing for the employment of an experimentally enhanced ozone regime indicates the canopy zone of "free-air" ozone fumigation (see text; for details, see Nunn et al. 2002; Werner and Fabian 2002; photo taken from the research crane by Dr. K.-H. Häberle)

and long-term environmental fluctuations (van Gardingen et al. 1997b). In this latter respect, the "quality of challenge" to some extent appears to be similar to that at the highest hierarchy levels in process scaling: Here, remote-sensing methodologies need to be adapted to ecological assessments, while the latter themselves need to be promoted into new research concepts in order to cope with the physical requirements and enhance the interpretability of remote-sensing in ecology (Ustin et al. 1993), i.e., making the quality of information obtained from both approaches compatible in a most proficient way.

8.3 Molecular Epidemiology

Plant diseases are an important limiting factor of plant growth, competitiveness, yield and reproduction. It is known, that plants whose photosynthesis is inhibited by some external stress factor often are disposed to an enhanced likelihood of plant disease. In addition, the uniform genetic structure of monocultures is known to facilitate the adaptation and attack of pathogens (Agrios 1997; Scheffer 1997). Interestingly, a thorough review of the reasons for epidemic plant diseases (Scheffer 1997) has concluded that usually anthropogenic factors are involved. Undisturbed natural ecosystems may possess a natural balance between plant species and pathogens, although "ebb and flow"-like local epidemics are common (Scheffer 1997). These general views are backed up only to a limited degree at the mechanistic and molecular levels.

Horizontal resistance that is polygenic and not specific for pathogen race is often more durable than vertical resistance based on gene-for-gene relationships (Brown 2002). The latter involves plant resistance genes often encoding leucine-rich repeat-type proteins (Hulbert et al. 2001) and pathogen avirulence genes whose products are often exported by type III transporters (Büttner and Bonas 2002). In spite of many advances concerning molecular mechanisms, it is still not known to what degree plant biodiversity is influenced by pathogens and to which extent pathogen-caused changes in biodiversity may quantitatively affect the resource fluxes and, hence, functional characteristics of ecosystems. This lack of knowledge is related to the fact that current plant pathology is strongly focused on crop plants (Agrios 1997; Scheffer 1997), although more than 90 % of carbon storage on earth is determined by woody plant systems and their effects on carbon uptake and retention in biomass and decomposition of the latter (Olson et al. 1983). Knowledge about the role of plant pathogens for wild plants, plant biodiversity and biogeochemical cycles is limited, although the recent intensive research on the weed plant species, *Arabidopsis thaliana*, has resulted in identification of molecular mechanisms resembling those of crop plants.

As pointed out above, it is still debated to what extent plant biodiversity influences resource fluxes of a given ecosystem, and how much pathogen impact can be tolerated prior to irreversible changes in ecosystem functioning. In cases such as chestnut blight or Dutch elm disease, the worldwide detrimental effect of pathogens is clearly documented. (Kranz 1996; Agrios 1997; Scheffer 1997). Still, it is not well known how an equilibrium between host plant and pathogens is maintained or lost in natural ecosystems, although a number of pioneering studies have been performed (Burdon and Leather 1990; Kranz 1996), and the concept of geophytopathology has been developed (Weltzien 1978; Kranz 1996). Stated in a general way, infection chains of pathogens, the nature of the non-host plant and the susceptibility as well as the vertical and horizontal resistance properties of host plants need to be determined in dependence on environmental parameters (Kranz 1996). Present concepts deal mainly with pathogen inoculum spread (e.g., Brown and Hovmoller 2002) and less with host defense status, which is conceptually known to depend on environmental conditions ("disease tetrahedron"; Agrios 1997). Biomarkers to judge the host status for pathogen susceptibility under conditions of air pollution or global climatic change are desirable, but still have to be established. A first analysis of worldwide *Arabidopsis* ecotypes for susceptibility to the bacterial pathogen *Pseudomonas syringae* surprisingly revealed a lack of correlation between disease susceptibility and plant fitness (Kover and Schaal 2002).

Molecular biomarkers are already well-established in medical epidemiology, such as in cancer research (Srivastava et al. 1994) and in the differentiation of genetic and environmental contributions to various human diseases (Schulte and Perera 1993). In an analogous manner, molecular biomarkers should be developed to characterize partial aspects of anthropogenic stress (Huggett et al. 1992) and in particular of plant fitness. For example, gene markers for systemic acquired resistance (SAR; Sticher et al. 1997), the hypersensitive response (see Chaps. 4, 5 and 6) and induced systemic resistance (ISR; see Chap. 8) have become available. Genome-wide microarray studies of SAR (Maleck et al. 2000) and HR (Scheideler et al. 2002) have been performed and have identified a surprisingly large number of response genes.

The use of biomarkers in plants is made difficult by the occurrence of delayed stress responses, in particular in tree species. This will be illustrated here by the controversy regarding the question whether elevated tropospheric ozone influences plant disease incidence. Large-scale US and European research programs have tried to clarify the effect of ozone on crop loss. However, these studies made use of pesticides so that effects of ozone on biotic disease incidence were suppressed (Sandermann 1996). The influence of the current high ozone levels, particularly on low-pesticide (e.g., organic) agriculture, remains undefined. Previous literature has implied that high levels of ozone and UV-B may not interact with plant disease, because "the two

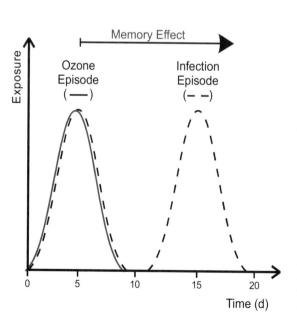

Fig. 8.6. Epidemiological scenario for ozone/biotic disease interactions. A summer period of high ozone (—) may coincide with increased attack by bark beetles (–; *left peak*) because both stresses are favored on sunny days. On the other hand, a fungal infection episode (–; *right peak*) is typically separated from high ozone days. This non-coincidence could suggest negligible ozone-biotic disease interaction. However, as indicated by the *arrow* labeled "*memory effect*," biochemical changes induced in a high ozone episode may persist long enough to be still present in a subsequent fungal infection period. (Taken from Sandermann 2000)

non-continuously acting factors ozone and UV-B radiation are more or less completely non-coincident with infection periods of most pathogens" (Manning and von Tiedemann 1995). This non-coincidence is generally true, but because of "memory effects" (Sandermann 1996), episodes of high tropospheric ozone do not have to coincide with the episodes of infection pressure. The delay periods of various molecular stress responses have been determined and found to reach from hours to well over a year. These results have led to the epidemiological research concept depicted in Fig. 8.6 (Sandermann 2000). The epidemiological significance of interactions between primary and secondary stressors is often neglected (Manion and Lachance 1992). For example, dead conifer trees in the German Fichtelgebirge (Schulze et al. 1989) and dead maple trees in Canada (Driscoll et al. 2001) were shown together with headlines on mineral deficiencies or acidic deposition. In these cases, the trees are known actually to be killed by biotic attack (Forschungsbeirat 1989; Eagar and Adams 1992; Manion and Lachance 1992; Heliövaara and Väisänen 1993), as also described for ozone-damaged *Pinus ponderosa* trees in the southern Californian San Bernardino Mountains (Miller and McBride 1998).

In summary, the great progress with regard to molecular mechanisms of plant disease or resistance is in contrast to the present lack of knowledge with regard to the environmental influence on disease incidence.

8.4 Connecting Laboratory to Field Experiments

Biomarkers will probably first be developed for agricultural crop plants that are much more genetically uniform than wild plant and forest tree species. For example, expression of the latent genes of SAR that are induced by plant strengthening agents such as Bion (Sticher et al. 1997) or by ozone (Sandermann et al. 1998) is detectable by array or biochip techniques. The study of various *Arabidopsis thaliana* ecotypes and wild relatives has generated valuable model systems for molecular ecology (Mitchell-Olds 2001). The short generation time of *Arabidopsis* has in one detailed case study allowed the demonstration of somatic and germline mutational changes as a result of slightly increased UV-B (Ries et al. 2000).

When the general topic of plant biodiversity is considered, ecotoxic effects on the individual plant have to be translated into population dynamics. For example, selection pressure by atrazine and many other herbicides has favored the predominance of resistant weed mutants in agroecosystems (Le Baron and Gressel 1982). Molecular herbicide resistance mechanisms of weeds have in many cases been linked to population dynamics (Powles and Holtum 1994). The broad-spectrum herbicide, glyphosate, has long been denied to lead to resistant weeds, but recently the development of resistant weeds by point mutation of the target gene product was acknowledged by the

Fig. 8.7. (in color at end of book) Aerial view and leaf area index of the experimental farm Scheyern near Munich, Germany. This figure was kindly provided by Dr. M. Sommer, GSF-Research Center, Institute for Biomathematics and Biometrics

manufacturing company (Baerson et al. 2002). The influence of high ozone has been shown to lead to changes in plant biodiversity in southern California (Miller and McBride 1998), but more generally it has been difficult to demonstrate population shifts by this stressor (Barker and Tingey 1992). A large variety of ecotypes of *Arabidopsis thaliana* are available, and comparative genomics may uncover traits determining host or non-host status, e.g., for the bacterial pathogen, *Pseudomonas syringae* (Kover and Schaal 2002).

The variability of field site conditions makes it usually quite difficult to establish causal relationships. Figure 8.7 shows a well-characterized agricultural field site (experimental farm Scheyern) that is located near Munich in Bavaria. The Scheyern farm area is particularly suitable for studying the functional effects of landscape heterogeneity, and. conventional agriculture with chemical inputs has been systematically compared to organic agriculture (Schröder et al. 2002). Like the above-mentioned Kranzberg forest site, the Scheyern farm is part of an experimental hierarchy that includes the laboratory level (e.g., use of cell cultures and of *Arabidopsis thaliana* accessions) and reproducible model ecosystems (e.g., exposure chambers, solar fields and lysimeters). Such graded experimental platforms seem generally necessary in order to scale up from molecular to ecological processes, as previously described for the topic of forest decline by ozone (Sandermann et al. 1997).

The combination of scaling-up and molecular research described in this book should help to correlate plant genotype with ecological phenotype. This could define some of the rules of plant evolution and, in extreme cases, extinction in the present stressful world.

References

Agrios GN (1997) Plant pathology, 4th edn. Academic Press, San Diego

Baerson SR, Rodriguez DJ, Tran M, Feng Y, Biest NA, Dill GM (2002) Glyphosate-resistant goosegrass. Identification of a mutation in the target enzyme 5-enolpyruvylshikimate-3-phosphate synthase. Plant Physiol 129:1265–1275

Baldocchi DD (1993) Scaling water vapor and carbon dioxide exchange from leaves to a canopy: rules and tools. In: Ehleringer JR, Field CB (eds) Scaling physiological processes, leaf to globe. Academic Press, San Diego, pp 77–114

Barker JR, Tingey DT (eds) (1992) Air pollution effects on biodiversity. Van Nostrand Reinhold, New York

Bazzaz FA (1997) Allocation of resources in plants: state of the science and critical questions. In: Bazzaz FA, Grace J (eds) Plant resource allocation. Academic Press, San Diego, pp 1–38

Bazzaz FA, Grace J (eds) Plant resource allocation. Academic Press, San Diego, pp 265–277

Bol JF, Linthorst HJM (1990) Plant pathogenesis-related proteins induced by virus infection. Annu Rev Phytopathol 28:113–138

Brown JKM (2002) Yield penalties of disease resistance in crops. Curr Opin Plant Biol 5:339–344

Brown JKM, Hovmoller MS (2002) Aerial dispersal of pathogens on the global and continental scales and ist impact on plant disease. Science 297:537–541

Büttner D, Bonas U (2002) Getting across-bacterial type III effector proteins on their way to the plant cell. EMBO J 21:5313–5322

Burdon JJ, Leather SR (eds) (1990) Pests, pathogens and plant communities. Blackwell, Oxford

Cao H, Baldini RL, Rahme LG (2001) Common mechanisms for pathogens of plants and animals. Annu Rev Phytopathol 39:259–284

Driscoll CT, Lawrence GB, Bulger AJ, Butler TJ, Cronan CS, Eagar C, Lambert KF, Likens GE, Stoddard JL, Weathers KC (2001) Acidic deposition in the Northeastern United States: sources and inputs, ecosystem effects and management strategies. BioScience 51:180–198

Eagar C, Adams MB (eds) (1992) Ecology and decline of *Red Spruce* in the eastern United States. Ecological studies 96. Springer, Berlin Heidelberg New York

Ehleringer JR, Field CB (1993) Scaling physiological processes – leaf to globe. Academic Press, San Diego, 387 pp

Ehleringer JR, Hall AE, Farquhar GD (1993) Stable isotopes and plant carbon-water relations. Academic Press, San Diego, 555 pp

Forschungsbeirat Waldschäden/Luftverunreinigungen (1989) Dritter Bericht. Kernforschungszentrum Karlsruhe GmbH, Karlsruhe

Gómez-Gómez L, Boller T (2002) Flagellin perception: a paradigm for innate immunity. Trends Plant Sci 7:251–256

Grace J, van Gardingen PR, Luan J (1997) Tackling large-scale problems by scaling-up. In: van Gardingen PR, Foody GM, Curran PJ (eds) Scaling-up from cell to landscape. Cambridge University Press, Cambridge, pp 7–16

Grams TEE, Kozovits AR, Reiter IM, Winkler JB, Sommerkorn M, Blaschke H, Häberle K-H, Matyssek R (2002) Quantifying competitiveness in woody plants. Plant Biol 4:153–158

Grime JP, Hodgson JG, Hunt R (1988) Comparative plant ecology. A functional approach to common British species. Unwin Hyman, London

Heil M, Baldwin JT (2002) Fitness costs of induced resistance: emerging experimental support for a slippery concept. Trends Plant Sci 7:61–67

Heliövaara K, Väisänen R (1993) Insects and pollution. CRC Press, Boca Raton

Herms DA, Mattson WJ (1992) The dilemma of plants: to grow or defend. Q Rev Biol 67:283B335

Huggett, RJ, Kimerle RA, Mehrle PM Jr, Bergman HL (eds) (1992) Biomarkers. Biochemical, physiological, and histological markers of anthropogenic stress. Lewis, Boca Raton

Hulbert SH, Webb CA, Smith SM, Sun Q (2001) Resistance gene complexes: evolution and utilization. Annu Rev Phytopathol 39:285–312

Janeway CA Jr, Medzhitov R (2002) Innate immune recognition. Annu Rev Immunol 20:197–216

Jarvis PG (1993) Prospects of bottom-up models. In: Ehleringer JR, Field CB (eds) Scaling physiological processes – leaf to globe. Academic Press, San Diego, pp 117–126

Kessler A, Baldwin IT (2002) Plant responses to insect herbivory: the emerging molecular analysis. Annu Rev Plant Biol 53:299–328

Kolb TE, Matyssek R (2001) Limitations and perspectives about scaling ozone impact in trees. Limitations and perspectives. Environ Pollut 115: 373–393

Kover PX, Schaal BA (2002) Genetic variation for disease resistance and tolerance among *Arabidopsis thaliana* accessions. Proc Natl Acad Sci USA 99:11270–11274

Kranz J 181996) Epidemiologie der Pflanzenkrankheiten. Verlag E Ulmer, Stuttgart

Kruijt B, Ongeri S, Jarvis PG (1997) Scaling of PAR absorption, photosynthesis, and transpiration from leaves to canopy. In: van Gardingen PR, Foody GM, Curran PJ (eds) Scaling-up from cell to landscape. Cambridge University Press, Cambridge, pp 79–104

Laurence JA, Amundson RG, Friend AL, Pell EJ, Temple PJ (1994) Allocation of carbon in plants under stress: an analysis of the ROPIS experiments. J Environ Qual 23:412–417

LeBaron HM, Gressel J (eds) (1982) Herbicide resistance in plants. Wiley, New York

Levin SA (1993) Concepts of scale at the local level. In: Ehleringer JR, Field CB (eds) Scaling physiological processes – leaf to globe. Academic Press, San Diego, pp 7–20

Maleck K, Levine A, Eulgem T, Morgan A, Schmid J, Lawton KA, Dangl JL, Dietrich RA (2000) The transcriptome of *Arabidopsis thaliana* during systemic acquired resistance. Nat Genet 26:403–410

Manion PD, Lachance D (eds) (1992) Forest decline concepts. APS Press, St Paul, MN

Manning WJ, von Tiedemann A (1995) Climate change: potential effects of increased atmospheric carbon dioxide (CO_2), ozone (O_3), and ultraviolet-B (UV-B) radiation on plant diseases. Environ Pollut 88:219–245

Matyssek R (2001a) How sensitive is birch to ozone ? Responses in structure and function. J For Sci 47:8–20

Matyssek R (2001b) Trends in forest tree physiological research: biotic and abiotic interactions. In: Huttunen S, Heikkilä H, Bucher J-B, Sundberg B, Jarvis PG, Matyssek R (eds) Trends in European forest tree physiological research. Kluwer, Dordrecht, pp 241–246

Matyssek R, Innes JL (1999) Ozone – a risk factor for trees and forests in Europe? Water Air Soil Pollut 116:199–226

Matyssek R, Sandermann H (2003) Impact of ozone on trees – an ecophysiological perspective. Progress in botany. Springer, Berlin Heidelberg New York, pp 349–404

Matyssek R, Reich PB, Oren R, Winner WE (1995) Response mechanisms of conifers to air pollutants. In: Smith WK, Hinckley TH (eds) Physiological ecology of coniferous forests. Physiological ecology series. Academic Press, New York, pp 255–308

Matyssek R, Günthardt-Goerg MS, Schmutz P, Saurer M, Landolt W, Bucher JB (1998) Response mechanisms of birch and poplar to air pollutants. J Sustainable For 6:3–22

Matyssek R, Schnyder H, Elstner E-F, Munch J-C, Pretzsch H, Sandermann H Jr (2002) Growth and parasite defense in plants; the balance between resource sequestration and retention. Plant Biol 4:133–136

Miller PR, McBride JR (eds) (1998) Oxidant air pollution impacts in the montane forests of southern California. Ecological studies 134. Springer, Berlin Heidelberg New York

Mitchell-Olds T (2001) *Arabidopsis thaliana* and its wild relatives: a model system for ecology and evolution. Trends Ecol Evol 16:693–700

Mooney HA, Winner WE (1991) Partitioning response of plants to stress. In: Mooney HA, Winner WE, Pell E (eds) Response of plants to multiple stresses. Academic Press, San Diego, pp 129–141

Müller I, Schmid B, Weiner J (2000) The effect of nutrient availability on biomass allocation patterns in 27 species of herbaceous plants. Perspectives in Plant Ecol Evol Syst 2/3: 115–127

National Research Council (2000) Genetically modified pest-protected plants. Science and regulation. National Academy Press, Washington, DC

National Research Council (2002) Environmental effects of transgenic plants. The scope and adequacy of regulation. National Academy Press, Washington, DC

Nunn AJ, Reiter IM, Häberle K-H, Werner H, Langebartels C, Sandermann H, Heerdt C, Fabian P, Matyssek R (2002) "Free-air" ozone canopy fumigation in an old-growth mixed forest: concept and observations in beech. Phyton (Austria) 42:105–119

Olson JS, Watts JA, Allison LJ (1983) Carbon in live vegetation of major world ecosystems. DOE/NBB-0037, Carbon Dioxide Division, US Dept. of Energy, Washington

Penmetsa RV, Cook DR (1997) A legume ethylene-insensitive mutant hyperinfected by its rhizobial symbiont. Science 275:527–530

Polle A, Matyssek R, Gunthardt-Goerg MS, Maurer S (2000) Defense strategies against ozone in trees: the role of nutrition. In: Agrawal SB, Agrawal M (eds), Environmental pollution and plant responses. Lewis Publishers, New York, Boca Raton, pp 223–245

Powles SB, Holtum JAM (eds) (1994) Herbicide resistance in plants. Biology and biochemistry. Lewis Publishers, Boca Raton

Reynolds JF, Hilbert DW, Kemp PR (1993) Scaling ecophysiology from the plant to the ecosystem: a conceptual framework. In: Ehleringer JR, Field CB (eds), Scaling physiological processes, leaf to globe. Academic Press, San Diego, pp 127–140

Ries G, Heller W, Puchta H, Sandermann H Jr, Seidlitz HK, Hohn B (2000) Elevated UV-B radiation reduces genome stability in plants. Nature 406:98–101

Sandermann H (1996) Ozone and plant health. Annu Rev Phytopathol 34:347–366

Sandermann H (1997) Transgene Pflanzen: ökologische Fragen. Spektrum Wissensch 7:38–41

Sandermann H (2000) Ozone/biotic disease interactions: molecular biomarkers as a new experimental tool. Environ Pollut 108:327–332

Sandermann H (2003) Flawed science underlies laws on transgenic crops. Nature 424:613

Sandermann H, Wellburn AR, Heath RL (eds) (1997) Forest decline and ozone. Ecological studies 127. Springer, Berlin Heidelberg New York

Sandermann H, Ernst D, Heller W, Langebartels C (1998) Ozone: an abiotic elicitor of plant defence reactions. Trends Plant Sci 3:47–50

Saxe H, Ellsworth DS, Heath J (1998) Tree and forest functioning in an enriched CO_2 atmosphere. New Phytol 139:395–436

Scheffer RP (1997) The nature of disease in plants. Cambridge University Press, Cambridge

Scheideler M, Schlaich NL, Fellenberg K, Beissbarth T, Hauser NC, Vingron M, Slusarenko AJ, Hoheisel JD (2002) Monitoring the switch from housekeeping to pathogen defense metabolism in *Arabidopsis thaliana* using cDNA arrays. J Biol Chem 277:10555–10561

Schenk PM, Kazan K, Wilson I, Anderson JP, Richmond T, Somerville SC, Manners JM (2000) Coordinated plant defense responses in *Arabidopsis* revealed by microarray analysis. Proc Natl Acad Sci USA 97:11655–11660

Schlesinger WH (1997) Biogeochemistry – an analysis of global change. Academic Press, San Diego, 588 pp

Schröder P, Huber B, Olazábal U, Kämmerer A, Munch JC (2002) Land use and sustainability: FAM research network on agroecosystems. Geoderma 105:155–166

Schulte PA, Perera FP (eds) (1993) Molecular epidemiology. Principles and practices. Academic Press, New York

Schulze ED (1994) Flux control in biological systems. Academic Press, San Diego, 494 pp

Schulze ED, Lange OL, Oren R (1989) Ecological studies 77. Forest decline in air pollution. Springer, Berlin Heidelberg New York

Smith TM, Shugart HH, Woodward FI (eds) (1978) Plant functional types. Cambridge University Press, Cambridge

Srivastava S, Lippman SM, Hong WK, Mulshine JL (eds) (1994) Early detection of cancer. Molecular markers. Futura Publishing Company Inc, Armonk, New York

Sticher L, Mauch-Mani B, Métraux JP (1997) Systemic acquired resistance. Annu Rev Phytopathol 35:235–270

Stitt M, Schulze E-D (1994) Plant growth, storage and resource allocation: from flux control in metabolic chain to the whole-plant level. In: Schulze ED (ed) Flux control in biological systems: from enzymes to populations and ecosystems. Academic Press, San Diego, pp 57–118

Tiedje JM, Colwell RK, Grossman YL, Hodson RE, Lenski RE, Mack RN, Regal PJ (1989) The planned introduction of genetically engineered organisms. Ecological considerations and recommendations. Ecology 70:298–315

Ustin SL, Smith MO, Adams JB (1993) Remote sensing of ecological processes: a strategy for developing and testing ecological models using spectral mixture analysis. In: Ehleringer JR, Field CB (eds) Scaling physiological processes – leaf to globe. Academic Press, San Diego, pp 341–358

van Gardingen PR, Foody GM, Curran PJ (eds) (1997a) Scaling-up from cell to landscape. Cambridge University Press, Cambridge, 386 pp

van Gardingen PR, Russell G, Foody GM, Curran PJ (1997b) Science of scaling: a perspective on future challenges. In: van Gardingen PR, Foody GM, Curran PJ (eds) Scaling-up from cell to landscape. Cambridge University Press, Cambridge, pp 371–378

Weltzien HC (1978) Geophytopathology. In: Horsfall JG, Cowling EB (eds) Plant disease, vol II. Academic Press, New York, pp 339–360

Werner H, Fabian P (2002) Free-air fumigation of mature trees. Environ Sci Pollut Res 9:117–121

Subject Index

Color Illustrations

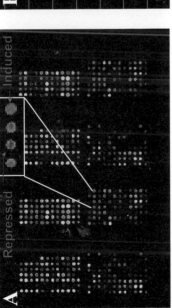

Fig. 1.8A, B. DNA microarray analysis of nitric oxide-induced gene expression in *Arabidopsis thaliana*. A Original scan of a DNA microarray hybridized with fluorescent-labeled samples (*green*, untreated control; *red*, sample from plants treated with nitric oxide, *NO*). *Green spots* indicate suppression, and *red spots* induction of the corresponding gene. B Global survey of transcriptional activity determined by a DNA array. The *scatter plot* shows significant gene activation after NO treatment (shift to the *right*). (Data from Jörg Durner, GSF-Forschungszentrum)

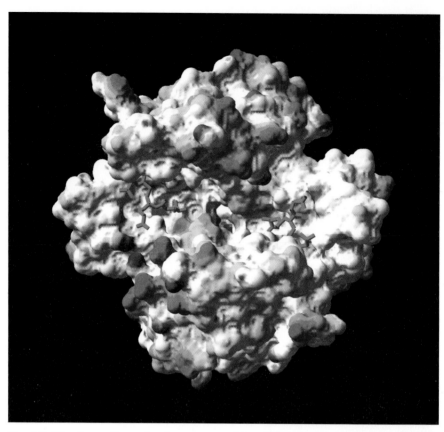

Fig. 2.2. The three-dimensional structure of *Zm*GSTF1–1, the major phi class GST expressed in maize foliage. The two *Zm*GSTF1 polypeptides are shown dimerised together, with the glutathione conjugate of the herbicide atrazine (shown in *green*) at the respective active sites. The structure represents the projected molecular surface and is coloured to represent the calculated electrostatic potential (*red* negative; *blue* positive)

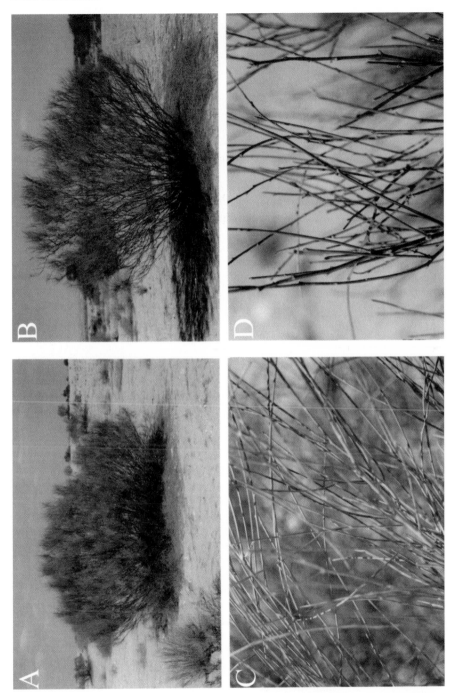

Fig. 3.2A–D. Dormant and non-dormant *Retama raetam* plants. **A** A non-dormant *R. raetam* plant (approximately 2 m high). **B** A dormant *R. raetam* plant (approximately 2.5 m high). **C** Stems of the non-dormant *R. raetam* plant shown in A. **D** Stems of the dormant *R. raetam* plant shown in B. The two plants were photographed at the same research site on the same day (in June 2000), about 20 min apart

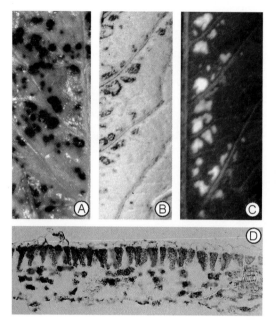

Fig. 4.3A–D. Colocalization of gene expression of the 'ethylene-forming enzyme' ACC oxidase, hydrogen peroxide accumulation and cell death in ozone-treated tomato plants (from Moeder et al. 2002). **A** GUS activity regulated by the Le-ACO1 promoter 1 h after the onset of ozone exposure. **B** H_2O_2 accumulation after 7 h. **C** Ozone-induced cell death after 24 h. **D** Transverse section through a leaflet showing GUS staining after 1 h

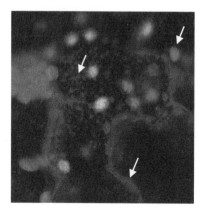

Fig. 5.1. In vivo imaging of a pathogen-induced NO burst in tobacco. NO production was visualized by confocal laser scanning microscopy of a cell treated with cryptogein, an elicitor from *Phythopthora* ssp. The cell was loaded with DAF-2DA, a specific fluorophore for NO. Six min after treatment with 50 nM cryptogein, DAF-2 DA fluorescence was localized along the plasma membrane and to distinct regions within the chloroplasts as well as to other subcellular structures, which may represent peroxisomes (*arrowheads*). Experimental details were as described previously. (Foissner et al. 2000)

Fig. 7.3. Reduction of Fusarium wilt in radish by biocontrol bacteria under commercial green-house conditions; *left plot* seeds treated with a coating containing *Pseudomonas fluorescens* strain WCS374; *middle plot* coating without bacteria; *right plot* non-treated seeds

Fig. 8.5. Experimental site "Kranzberg Forest" (near Munich, Germany) for studying mechanisms of competitiveness and ozone sensitivity in adult beech and spruce trees. Vertical tubing for the employment of an experimentally enhanced ozone regime indicates the canopy zone of "free-air" ozone fumigation (see text; for details, see Nunn et al. 2002; Werner and Fabian 2002; photo taken from the research crane by Dr. K.-H. Häberle)

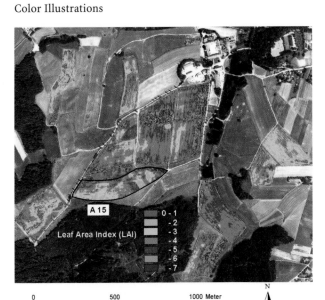

Fig. 8.7. Aerial view
and leaf area index of
the experimental farm
Scheyern near Munich,
Germany. This figure
was kindly provided by
Dr. M. Sommer, GSF-
Research Center, Insti-
tute for Biomathemat-
ics and Biometrics

Ecological Studies
Volumes published since 1998

DATE DUE